The Gene Revolution and Global Food Security

Also by Banji Oyelaran-Oyeyinka and Padmashree Gehl Sampath
LATECOMER DEVELOPMENT: Innovation and Knowledge for Economic Catch-up

Also by Banji Oyelaran-Oyeyinka
UNEVEN PATHS OF DEVELOPMENT (*with Rajah Rasiah*)

INDUSTRIAL CLUSTERS AND INNOVATION SYSTEMS IN AFRICA:
Institutions, Markets and Policy (*with Dorothy McCormick*)

SMES AND NEW TECHNOLOGIES: Learning E-Business and Development
(*with Kaushalesh Lal*)

LEARNING TO COMPETE: Institutions, Technology and Enterprise in Africa

Also by Padmashree Gehl Sampath
UNHEALTHY DIVIDE: Local Capacity for Disease of the Poor

REGULATING BIOPROSPECTING: Institutions for Access and Drug Research

The Gene Revolution and Global Food Security

Biotechnology Innovation in Latecomers

Banji Oyelaran-Oyeyinka
Director, Monitoring and Research Division, UN-HABITAT

Padmashree Gehl Sampath
Research Fellow, United Nations University-MERIT

First published 2009 by
PALGRAVE MACMILLAN

Palgrave Macmillan in the UK is an imprint of Macmillan Publishers Limited, registered in England, company number 785998, of Houndmills, Basingstoke, Hampshire RG21 6XS.

Palgrave Macmillan in the US is a division of St Martin's Press LLC, 175 Fifth Avenue, New York, NY 10010.

Palgrave Macmillan is the global academic imprint of the above companies and has companies and representatives throughout the world.

Palgrave® and Macmillan® are registered trademarks in the United States, the United Kingdom, Europe and other countries
ISBN-13: 978-0-230-22882-5   hardback

A catalogue record for this book is available from the British Library.

A catalog record for this book is available from the Library of Congress.

10 9 8 7 6 5 4 3 2 1
18 17 16 15 14 13 12 11 10 09

For
Fola, Banke, Koye
and
Nisha

Contents

List of Boxes

List of Figures

List of Tables

Preface

This book is set against the backdrop of the debate about the persistent poverty of developing countries – those we have differentially labelled *latecomer countries* – as well as the widening divide in income per capita rather than the convergence of income that was predicted several decades ago. Equally germane to the debate about latecomer poverty and growing inequality between the rich and the poor countries is the ever-widening gap between the scientific and technological capabilities of latecomer countries and the 'frontier' advanced industrial nations. Underlying these debates is considerable confusion about what precisely the role of states may be in the economic advancement of those countries that have developed and the enduring poverty of those that remain at the bottom of the ladder. This book argues that those who advocate a neutral role for state industrial and innovation policy tend to engage in rhetoric that does not fit the facts of the history of Western societies and simply ignores the contemporary lessons of East Asia. The current triune global challenge, namely, the food crisis, fuel and climate concerns, and the housing crisis has renewed the discussion about the form, nature, and direction of participation in business and society.

This book benefits from our collective involvement in the field of industrial, scientific, and technological policy for more than three decades both as practitioners and as researchers. It is in fact our experiences in different countries in Africa and Asia working, researching, and interviewing all manner of actors such as large-scale farmers, industrialists, and small enterprise owners that prompted this book as well as other companion books in which we have been engaged. The research and outputs thereby suggested to us to seek – with others – that the most urgent issue for these times is to formulate some answers and possible prescriptions to the persistent latecomer economic backwardness.

Beginning with the concepts of the processes we call variously 'technology transfer', 'technology capability acquisition', and so forth, from the 1970s the notion of innovation entered the lexicon of the process of development now referred to as 'catching-up' by latecomers. In the six years we spent at the UNU-INTECH, we led several research projects (interviewing dozens of firms) in attempting to understand the root of technology-based and industry-based underdevelopment. This endeavour led to several books and articles in academic journals, including one recently published and another forthcoming book.[1]

These experiences led us to three sets of conclusion: first, rich countries built strong institutions as complements to productive systems (frontier as well as emerging), and in so doing, became rich through production and exporting of high quality goods and services; poor countries remain poor because they continue to produce raw materials for the relatively rich countries. Second, central to the production activities of all countries that became rich is a set of policies that we may classify as industrial or innovation policies even when the rhetoric of rich countries does not square with the reality as exemplified by the set of ideas embodied in the so-called Washington Consensus (WC). In other words, rich countries do precisely what benefits their industries at every historical turn and continue to advise poor countries to do what benefits rich countries. The 'strong' poor countries (India, China, and much of East Asia earlier on) deployed Industrial and Innovation Policies (IIP) for the benefit their economies while the 'weak' poor countries (much of sub-Saharan Africa (SSA) and the other Asian countries) listen to variants of advice embodied in the WC and subsequently remain poor. This variant of the WC was precisely what the Structural Adjustment Programme (SAP) was about in the 1980s. This set of 'one size fits all' policies replaced many of the (admittedly more complex) emerging industrial policies in SSA with a simplistic macro-economic framework that led to the de-industrialization of the embryonic economies of SSA. The third broad conclusion is that poor countries require industrial and innovation policies that shift productive attention from raw materials production to materials processing using the best technologies (in economic terms, structural transformation for turning out high-quality products) available. These technologies also result in a rise in labour productivity (rising output per worker). We have defined what we understand by these terms in the book and, more importantly, we believe the set of ideas guiding what we term innovation capacity should be linked to the traditional issues of development economics because solving the poverty problem engendered by the endemic low productivity regimes of production in very latecomers will have to reckon with building innovation capacity underpinned by context-specific institutions and policies.

We argue that latecomer countries need to build innovation capacity for three key reasons, namely: to promote strong interactive learning among productive and non-productive actors; to foster greater information and knowledge flow, and, lastly, to enhance the coordination of policies and actors. Undertaking these three is a complex task and this has been the usual argument against industrial policy: that is, governments are unable to allocate resources through bureaucratic mechanisms and are in any case too corrupt and lack the incentive to work towards public purpose. In short, advocates of the alternative systems argue that IIP is too complicated and complex for latecomers in Asia and Africa to engage in. Two sets of issue raised by critics of IIP are: first, a lack of incentive for good bureaucratic behaviour, and second, the state in latecomer countries is unable to substitute for information coordination due to its highly decentralized nature. We consider this argument unconvincing because the same case can be made to support the establishment of mechanisms, organizations, and institutions for education, health, roads, and so on in the form of ministries and agencies.

The plethora of myths about markets as well as new and important evidence on the role of the state reinforce the need for a rethink of what the role of the government should be in creating dynamic innovation policies in a global system that has become more complex, knowledge-based, and innovation-driven. These are the ideas that we put forward in making a case for a redoubling of efforts in latecomer countries to deploy biotechnology in solving both the poverty problem and the food crisis in poor countries.

Acknowledgements

The idea of embarking on a research project to compare the biotechnology systems of innovation in African and Asian countries arose out of some of our earlier work on the topic, as well as the curiosity to find new data on issues so pertinent to sustainable development among latecomers. We are grateful to IDRC, Canada for funding the project which led to the research results contained in this book. Field surveys in latecomer countries on topics such as these are impossible and incomplete without the assistance of committed local researchers. We acknowledge and thank all our consultants in Vietnam, Malaysia, Nigeria, Tanzania, Ghana, and Kenya who helped conduct the empirical surveys. Our special thanks go to Wladimir Raymond, Erika Moran, Graciela van der Poel, Eveline in the Braek from UNU-MERIT, and Nelly Kangethe from UNHABITAT for research assistance at different stages of book production. Last but not least, we thank Rosemund and Dirk for their never-ending support and patience, especially in accommodating the hectic travel schedules that this book necessitated.

1
Agricultural Biotechnology Innovation Capacity and Economic Development

1.1 Introduction

This book presents the findings of a multi-year study of the agricultural system of biotechnology across six countries in Asia and Africa. The basic proposition of the study that informs the book was applied across all countries. The key questions were framed to expose what those countries in the catch-up phase require to build and sustain a competitive science and technological infrastructure to deal with the food crisis and to solve the underlying challenge of poverty. The underlying hypothesis of this line of inquiry is that the resolution of the problems of endemic poverty will require – among other development efforts – that countries make long-term sustainable investments not only in science and technological infrastructure but also in developing the right kinds of institution and policy to exploit modern biotechnology. To achieve this, *latecomer countries* will have to invest in resources for building a complex multidimensional and dynamic range of knowledge, skills, actors, institutions, and policies within specific political policy structures defined as 'innovation capacity'.

The success of the Green Revolution in much of Asia and the lack of transformation of Africa's agriculture in spite of research efforts over the last few decades is increasingly being explained in the context of changing knowledge and capabilities of countries. Agricultural development depends to a great extent on how successfully knowledge *of relevance to the local context* is generated and applied. Investments in knowledge – especially in the form of science and technology – have featured prominently and consistently in most strategies to promote sustainable and equitable agricultural development at the national level. Although many of these investments have been fairly successful, the context for agriculture is changing rapidly – sometimes radically – and the process of knowledge generation and use, and agricultural innovation, has also been transformed.

The Green Revolution that helped achieve agricultural self-sufficiency in several parts of the world was characterized by state-led, public-sector-driven research on varieties of rice and wheat, a focus on technological infrastructure that increased land productivity, physical infrastructure that enhanced proximity to markets and, most importantly, high yielding crops. Whereas it can be argued that the lack of a state-led vision of the kind observed in India, Mexico, and other countries that pioneered the Green Revolution stifled African success in this regard, several other changes mark the emerging global agricultural landscape. Since the 1980s private-sector organizations have taken over much of the task of providing high yielding varieties for agricultural production worldwide, which was the product of public-sector research during the Green Revolution. Much of the focus of the private sector continues to be on rice and wheat (which were the main varieties of the Green Revolution) apart from maize, cotton, soybean, and barley. Significantly, local varieties of importance to Africa, such as cassava, sorghum, and banana have not been the focus of private-sector agricultural research.

Five changes in the context of agricultural development call attention to the need to examine how innovation that underpins greater productivity occurs in the agricultural sector: (i) markets, not production alone, increasingly drive agricultural development; (ii) the production, trade, and consumption environment for agriculture and agricultural products is dynamic and is evolving in unpredictable ways; (iii) knowledge, information, and technology of relevance to agriculture are increasingly generated, diffused, and applied through the private sector; (iv) exponential growth in information and communications technology (ICT), especially the Internet, has transformed the ability to take advantage of knowledge developed in other places or for other purposes, and (v) the knowledge structure of the agricultural sector in many countries is changing markedly.

These factors have changed the face of agricultural development and rendered it intricately linked to global economic trade and to the knowledge capabilities of countries. But they also, more than ever before, focused the associated emphasis on the inability of economic growth to address the food security needs of the poor. Placing agriculture in a knowledge-based innovation-driven context points attention to the notion of science, technology, and innovation capacity-building and what the prospects might be for achieving a Green Revolution in latecomer countries, especially those in sub-Saharan Africa. How can latecomers achieve their own Green Revolution? What are the main points of trigger, both state and market-oriented, that will help focus the agricultural system of production and innovation on local needs? In finding answers to these questions, this book makes three key proposals: first, that science, technology, and innovation capacity should be central and not peripheral to the developing of agricultural biotechnology systems of innovation in Africa. Second, we posit that discussions about food security in latecomer countries should emphasise on what is now termed 'innovation policy' rather than just science, technology, or research and development (R&D) policy. The differences in these concepts are explained here and expounded upon further in the subsequent chapters of the book. For instance, we cannot divorce innovation policy from educational policy because the latter provides the vital scientific and technical manpower for research and production. The need to develop innovation capacity is urgent because of the difficult overall environment of the innovation system[1] in latecomers, which is poorly organized and somewhat fragmented. In rural as well as urban settings a large number of farm and non-farm actors operate a subsistence and informal economy. The few foreign-based firms tend to be disconnected from the rest of the economy. In the knowledge sector, the research community, which includes the universities, usually operates in an ivory tower, and is therefore poorly connected to local realities, particularly to labour market needs and opportunities. Particularly problematic are the lack of technological support services and infrastructure (e.g., metrology, quality control, and standards).

There are a number of public-sector organizations but with doubtful effectiveness, including those supporting the promotion of enterprise development, export, and foreign investment. In this often over-crowded support system, it is not easy to establish new, efficient organizations for the promotion of innovation. Where this is possible, the organizations are rarely appropriate and lack the flexibility and drive crucial for entrepreneurship. These overall conditions keep innovation systems in a low equilibrium trap. They are characterized by low levels of R&D in the business sector; the bulk of national R&D effort is borne by the government, and with questionable relevance for the economy. When research programmes are purely supply-driven, local market considerations tend to take second place and as a result 'commercialization' efforts then read like an afterthought.

Importantly, this book focuses on what might well be the capstone of an innovation-based approach to development which is the imperative of developing systemic linkages between public and private actors and investments; an idea that places the state at the forefront of coordinative interaction. For instance, efforts to raise the agricultural production and competitiveness are directly linked to the development of the agricultural value chain which includes small and medium agro-industrial enterprises as much as large conglomerates. Developing science, technology, and innovation (STI) capacity means that the internal capabilities of enterprises require development. How and what combination of these enterprises a country develops in the short and long term is not accidental and will be a result of historical path dependence as well as the nature of institutions and policies that the state is able to promote. In other words, a well-articulated vision of a long-run agricultural programme is an urgent imperative.

And there is clear evidence of the immense opportunities and scope for latecomers' own Green Revolution (LGR), especially for countries in Africa. There has been a large literature that tried to explain the failure of Africa's agriculture to benefit as much as Asia did from crop varietal improvements ranging from institutional and policy failures to geographic and historical path-dependent reasons. One of the central arguments of this book is that *the challenge of Africa's Green Revolution (AGR) needs to be examined from institutional systemic perspective and how factors connected to this promote or retard knowledge creation and diffusion.* According to Evenson and Gollin (2003), 'the cropping mix and *inherited state of knowledge* (and of germplasm) are the dominant factors in differential regional performance. Clearly there are institutional and political failures in all regions'. Herein lies the opportunity for sub-Saharan Africa that, due to the possibility to purposively raise investment in R&D and to improve the state of knowledge facilities, varietal improvements could be pursued with vigour based on extant and future created knowledge.

1.2 Agricultural innovation and the promise of biotechnology

Despite earlier misconceptions about the contribution of agriculture to the economy, by the mid-1960s this notion had been completely supplanted. This coincided with the fundamental changes made to agriculture by the innovation-driven Green Revolution. As in the case of many of the learning processes that characterized industry, considerable emphasis was placed on transfer of technologies from developed to developing countries.

The notion advanced by Theodore W. Schultz (1964), that viewed peasant producers as 'efficient but poor' (Abler and Sukhatme, 2006) pointed to the need to create conditions in resource-poor environments in which agriculture will be made more efficient based on productivity enhancing technologies. While the factors of nature – particularly land and water – were basic requirements for sustained agricultural production, new technologies and innovative institutional arrangements would be essential to ensure sustainability and security.

Drawing on earlier body of work, Mosher (1966) proposed five 'essentials' factors for farmers for agriculture development.[2] They are: (1) markets for farm products – fully complemented with infrastructure to transport products and the establishment of agro-enterprises to build and sustain the food supply chain; (2) continuous changes to technologies (read innovation) that raises productivity; (3) domestic supply of machinery and equipment – which speaks to local agricultural machinery and components development; (4) production incentives for farmers – an issue that points to the role of the correct policies for the agricultural sector including, and not limited to, fair prices for farm products, and (5) transportation – which anticipates the need for broader forms of infrastructure, including extension services.

Clearly the nature of innovation requires new actors in the form of educated scientists who would be the carriers of new knowledge to replace the largely fragmented agrarian knowledge system that forms the basis of rural poverty. The subsequent emphasis on greater educational investments and the establishment of scientific and technological infrastructure form the basis of the knowledge-based innovation approach in this book. Both the Green Revolution and its successor, the 'Gene' Revolution (that employs molecular and recombinant Deoxyribonucleic Acid (DNA) technologies and biotechnologies to enhance yields) demonstrates the gradual convergence of knowledge capabilities; now a global reality. Patterns of knowledge change are coming to be related to the increasing convergence in the different areas of science and technology. The Gene Revolution, for instance, is the result of the convergence of techniques and practices that encompass genomics, molecular biotechnologies, agricultural and industrial biotechnology within the biological sciences and biotechnologies knowledge base. Not only does this call attention to the state of knowledge infrastructure in latecomers to make the Green Revolution happen, it also calls forth institutional capabilities that will be needed to tailor appropriate institutional innovations to mitigate the divisive role played by Green Revolution packages in the frontier countries where agricultural success was concentrated among larger, resource-rich farmers, thus exacerbating social inequalities and land-tenure issues that still exist in these countries (Thomson et al., 2007).

For example, the world's poorest farmers account for 60% of total global agriculture and 80% of all agriculture in developing countries, but manage to produce only around 15 to 20% of world's food produce (see for example, Spillane, 2002). A large part of this 60% are African farmers, and have been classified as 'resource poor', which largely refers to fragmented, low input-oriented farming that is characterized by low soil fertility, extreme climate considerations, equity and land-tenure issues, and lack of access to competitive markets for both agricultural inputs and outputs.

Poverty that affects large swathes of the populations of the very late-comer countries is not only widespread but persistent over time. For instance, whereas both China and India have succeeded in reducing poverty over the last quarter century, the proportion of the population living in poverty remains large while urban populations continue to rise proportionally (see figures 1.1 and 1.2). In other words, the poorer countries will have to cope not just with increasing population pressures but an urban population with a greater awareness of basic rights, albeit, poor and largely uninvolved with food production.

Apart from rainfall levels, other extraneous influences caused by biotic and abiotic stresses give rise to particular forms of viruses or enhanced rates of soil erosion. These affect farming and yield rates. Therefore, a critical issue in promoting economic catch-up in agricultural development remains one of transposing success through lessons learned from other contexts into local realities in a sustainable and desirable way. These are addressed in subsequent sections and chapters.

1.3   Defining biotechnology and articulating agriculture biotechnology

Biotechnology is often regarded as the 'application of biological organisms, systems, and processes to manufacturing or service industries'. The lack of a widely and commonly accepted definition of biotechnology often renders studies seeking to study its application very difficult. Keeping this in mind, this book accepts the view that biotechnology comprises a continuum of technologies ranging from traditional to modern hence any technique that uses living organisms, or substances from those organisms, to make or modify a product, to improve plants or animals, or to develop micro-organisms for creation of new knowledge (see for example, OTA, 1989). We proceed with the awareness that biotechnology is not a sector but rather a set of techniques. Also that the sectoral characterestics in the analysis are therefore those of the agricultural sector, wherein the complementarity of biotechnology-based innovation play a major role due to its potential to improve crop yield and reduce poverty.

However, a profitable exploitation of the benefits of biotechnology is based on the scientific and technological understanding of the nature and characteristics of biological organisms and this function at the molecular level. This understanding and capability-building includes techniques for the manipulation of the DNA molecules – the source of the genes from which new organisms are made – in order to achieve healthier traits (Graff et al., 2005). The agricultural biotechnology techniques that this book and its analysis considers are (OECD, 2002, p. 30):

(a) DNA-based, including genomics, pharmacogenetics, gene probes, DNA sequencing/synthesis/amplification, and genetic engineering and genetic modification techniques;
(b) Proteins and molecules-based techniques, including protein sequencing, proteomics, hormones and growth factors research, and cell receptors/signaling/pheromones;
(c) Cell and tissue culture and engineering, including tissue engineering, marker-assisted selections, cellular fusion, hybridization, embryo manipulation, and vaccine/immune stimulants;
(d) Process biotechnology, including bioreactors, fermentation, and bioprocessing.

The emergence of these modern biotechnology techniques since the early 1980s has led to an increasing knowledge of the scientific procedures to utilize gene-based techniques to improve agriculture. However, these have mostly been in the forte (and developed by) commercial agriculture propped adequately by the benefits conferred in the form of intellectual property rights on biotechnology (breeders rights largely, but also patents in some countries). These gains, both in terms of agricultural products as well as intellectual property, have been uneven and skewed in the main by levels of capabilities, and the innovation capacity that a country is able to deploy at several levels from research to commercial application. Box 1.1 shows some of the contributions of science and technology to agricultural improvements.

Box 1.1 Past contributions of science and technology

The historical focus of research on food-crop technologies, especially genetic improvement of food crops, has undeniably been successful. Average crop yields in developing countries have increased by 71% since 1961, while average grain yields have doubled (to 2.8 tons per hectare). Yields of many commercial crops and livestock have also grown rapidly (see figure 1.1). The International Food Policy Research Institute (IFPRI) studies on impacts of public investment in India and China showed that agricultural R&D had higher impacts on poverty reduction compared to most other public investments, second only to the investment in education in China, and to rural roads in India (Fan, Zhang, and Zhang, 2000; Fan, Hazell, and Thorat, 1999). Other studies have shown that a 1 percent increase in agricultural yields in low-income countries leads to a 0.8% reduction in the number of people below the poverty line (Thirtle, Lin, and Piesse, 2003).

Source: World Bank 2006, FAOSTAT 2002 (for figure).

Graff et al. (2005) identify three broad directions deriving from traditional breeding methods which the gene techniques could take as useful entry points pointing to the potential for latecomer countries to build on extant capabilities. These possibilities, however, depend on these countries making the right kinds of investment in human capabilities and laboratory facilities within their own institutional contexts. The traditional breeding techniques could transform into the Gene Revolution II, if the threshold for convergence of new technologies is reached[3] through appropriate actions. The first is the acquisition of capabilities in genetic engineering, which is deployed to broaden the range of new products with better traits in both animals and plants. The relevant techniques include recombinant DNA techniques, insertion of genetic materials of plants and animals: wide crossings (genes transfers from wild relatives of the crop) and transfers of foreign genes (gene transfers across species). Second, deepen and increase the pace of plant breeding by using techniques such as selectable gene markers, promoters, and new scanning devices. The third and final one is the influence of policy and institutional changes that lead to cheaper processes for making products and collaborative and partnership strategies for raising systemic gains. For the relatively poor latecomer countries, genetic modification (GM) technology offers the possibilities of raising the productivity of small farmers across a wide swathe of agro-ecological zones employing gene transfers to introduce desirable traits that these plants lack into the best plant varieties.

Agricultural biotechnology has the potential to attenuate much of the widespread global poverty by raising agricultural productivity and raising farmers' wages through increasing crop yields. The recent global food crisis makes this more urgent for latecomers to adopt modern techniques of agricultural biotechnology. According to the Food Price Index compiled by the Food and Agriculture Organization (FAO), worldwide food prices have exhibited a continuous upward trend since 2002. Apart from sugar and meat, distinct increases were recorded for oils and fats, cereals, and diary, as reported in Figure 1.1. A similar uptrend was also observed for the International Monetary Fund's (IMF's) index of internationally traded food commodities prices, with an increase of 130% from 2002 to mid-2008, driven mainly by grains, and fats and oils[4] (Mitchell, 2008). While the upward trend abated somewhat from the mid-2008, food prices in 2008 were still more than 50% higher than that in 2002. While the impact of high food prices will vary across households in different countries, there is increasing evidence that the poor sections of the population in latecomer countries are particularly vulnerable (FAO, 2008; Ivanic & Martin, 2008). While farmers in the rural areas may benefit from a higher selling price of their produce, the urban poor spend more than half of their income on food and equally on other services and will be rendered far more vulnerable.

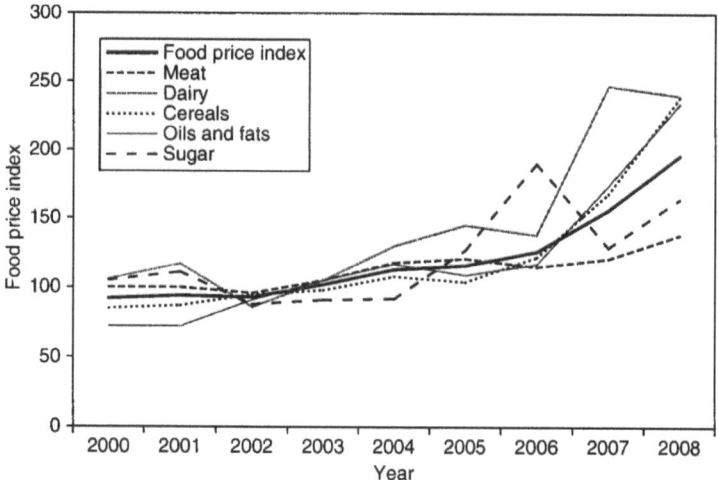

Figure 1.1 FAO food Price Index, 2000–8

The potential benefits of agriculture biotechnology for poverty reduction therefore include: yield increases of staple food crops that are produced in tropical and semi-tropical environments presently using low productivity technologies, See Box 1.2. Another benefit is the creation of drought- and pest-resistant varieties that can be propagated in an expanded areas of land and in problem-ridden areas where such activities were previously impossible. Biotechnology fosters multiple cropping and enables shorter harvesting cycles thereby increasing the possibility of planting several crops per season; and finally, the enablement of cost-saving techniques such as nitrogen fixation.

Box 1.2 Biotechnology, food security, and poverty

The world's population has grown nearly four-fold over the last century and is projected to rise from more than 6.6 billion people today to more than 8 billion by 2030. In the US, population has tripled over the last century. Since 2000 alone, the population of the US has grown by 20 million people, more than the population of the state of New York. At the same time, the world's hungry and chronically malnourished totals *830 million people*, despite global pledges and international efforts to improve food security.

Already biotechnology is preventing the loss of billions of pounds of important commodity crops such as corn and soybeans, and is expected to make an even larger yield contribution in the future.

Crops improved through biotechnology are increasing yields worldwide. Higher-yielding crops can help feed more people and boost incomes for poor farmers. For instance:

In South Africa, large and small-scale farmers have adopted biotech maize, soybeans, and cotton, which had contributed to an estimate US$156 million increase in farm income from 1998 to 2006.

Biotech cotton (resistant to the often-devastating bollworm insect) raised yields by 29% in India, and contributed to a 78% increase in income for many of the country's poorest farmers.

Enhanced varieties of corn have boosted yields worldwide – by as much as 61% over traditional varieties in the Philippines, where the average income for biotech corn farmers has increased by 34%.

The use of biotech crops that resist pests and diseases, tolerate harsh growing conditions, and reduce spoilage has also prevented the loss of billions of pounds of important crops.

In the US, enhanced crops have helped farmers prevent the loss of approximately 8 billion pounds of crops in 2005, according to experts.

Diseases and pests reduce global production of food by more than 35% – a cost estimated at more than US$200 billion a year. Scientists continue to work to develop a new generation of biotech crops to address these challenges, to do more to increase the yield of commodity crops and to help plants use water and nitrogen more efficiently.

Biotechnology has also contributed to improvements in crop productivity – helping plants become more efficient – and has the potential to increase productivity by another 25% worldwide. This can be achieved on existing farmland, to meet local needs in both developed and developing countries, where predictable and stable food production is particularly important.

Crops improved by biotechnology are embraced by farmers around the world. Over 12 million farmers in 23 countries – more than 90% of whom are resource-poor farmers in the developing world – are already planting biotech crops.

Agricultural biotechnology holds enormous promise for helping poor people around the world. Today, many farmers in the developing world choose biotech crops to boost productivity and increase efficiency – as a way of helping reduce poverty.

The benefits of biotechnology are passed on through a seed or plant cutting, so that farmers anywhere around the world are able share in the technology. That is why biotechnology is particularly attractive to scientists and rural development experts in poor countries where most people farm for a living.

The next generation of biotech crops is being developed to do even more to increase the yield of commodity crops, and to help plants use water and nitrogen more efficiently.

In addition to yield and productivity improvements, scientists are investigating how to use biotechnology to improve the nutritional profile of crops eaten by the poor.

Agricultural biotechnology also offers the possibility of developing varieties and techniques that reduce the use of fertilizers and pesticides while at the same time conserving other natural resources such as soil, water, and genetic diversity (Cook, 1993). For example, herbicide-resistant crop varieties can – in some cases – contribute to a dramatic reduction in pesticide usage (Fernandez-Cornejo and Mcbride, 2000). Some of these crops allow farmers to use more benign herbicides than the more harmful ones that were in use earlier. For example, Glyphosate, one such herbicide presently being used with GM crops is considered to be totally environmentally benign. Agricultural biotechnology can also help create new cultivars with increased resistance to other biotic stresses increasing the possibility of farming with lesser inputs such as water and energy. This squares well with the many of the natural resource-limited environments in latecomer countries (Gehl Sampath and Tarasofsky, 2002).

In sum, the trajectory of the first Green Revolution is being projected into the *Gene* Revolution II (from 1975 to 2009) a phenomenon made possible only by embedding the traditional systems in new scientific systems with important implications for new actor configurations, new institutions, and policies. The challenge for latecomers is clearly not just technical, but more poignantly, institutional as countries face the urgent tasks of reducing poverty and raising wages in a knowledge-based and innovation-driven global environment.

1.4   Building agricultural biotech capacity for food security and poverty reduction

In the previous section, we pointed out that agricultural biotechnology holds considerable promise for poverty reduction in wide variety of ways. Improving the livelihood of millions in latecomer countries will involve improving the productivity of the farming system, adapting and not necessarily abandoning the old traditional breeding techniques to assist smallholders in developing countries. In doing this, five broad concerns should be addressed, namely: building innovation capacity, fostering collaborative interactive learning, encouraging the emergence of new actors (scientists, breeders, etc); engaging policies that promote agriculture biotechnology and, finally, promoting institutions and institutional structures in given contexts.

Mapping the knowledge base within a context marked by dynamic agricultural system changes, assessing the competence of actors within these systems, and benchmarking them against global practices are essential inputs in the design of innovation policies. From this perspective, the system of innovation approach is a powerful tool for national and local policy-making. It provides a new way to organize knowledge as an input into policy-making, a means to analyse the support structures and policies needed for innovation, and a framework for situating the local in the context of dynamic processes of change at the global level. This is particularly important in designing policies to support interactive learning and innovation in agricultural systems.

In our conceptualization, we move beyond the simple quasi-autonomous process of *learning by doing*, to a more nuanced notion of 'learning to learn', a conscious process in the absence of which firms and farms neither improve productivity nor develop the capacity to innovate in products or processes as competitive conditions change. The linkages that different actors (smallholder farmers for instance) establish with local research institutions as well as commercial clients and suppliers at home and abroad can be critical in this respect. Moreover, the accelerated pace of technological change requires a far larger volume and set of resources than small, medium, and even large farmers traditionally have in-house or can easily access in latecomer countries. These resources are both knowledge-based and financial. Small producers, in particular, lack buffers that reduce the risks of change. They possess limited scanning capabilities because they are small and have equally limited financial resources because banks and capital markets are not major providers to this sector. For example, crop export capabilities, as other knowledge capabilities, involve considerable explicit efforts in mastering the techniques of long-distance handling and transport of fresh produce. Building a system involves diversification, a process that demands linkages in several directions. For instance, upstream linkage fosters interaction and learning from the capital goods sub-system and input suppliers. Downstream linkage promotes links with auxiliary producers and greater value added in the final products and services market. The classical examples are the tremendous opportunities that exist within the foods, beverages, and agricultural raw materials such as cotton and coffee.

Learning, however, is not solely the province of an isolated actor such as a firm or a farm. System-wide learning as we have characterized supports micro-level learning and is a prerequisite for keeping up. Knowledge gained in carrying out a set of tasks can be broadened and reused in new activities. The knowledge of enzymes acquired in mineral bleaching can be developed for a variety of purposes such as environmental clean-up. Enzymes are also important agents in foods and beverages for example in the conversion of cereals into malt in the brewing industry. A systems view of such knowledge gained across 'subsystems' opens important windows of opportunity to deepen local linkages. Traditional sectors therefore provide a useful platform for latecomer catch-up but policymakers need to reconceptualize these sectors as innovation systems for their dynamic potential to become more evident.

Comparative advantage in agricultural resources by itself is a static condition that can no longer form the basis of competitiveness. Regarded as a platform from which dynamic conditions can emerge, agricultural crops, for instance, can become a platform for catch-up. However, these need to be supported through technological improvements that render local production competitive. New technologies such as biotechnology have revolutionized traditional sectors from mining, agriculture, and fishing to services. The old notion that divides sectors into 'hi-tech' and 'low-tech' based strictly on R&D intensity is misleading viewed against the progressive intensification of knowledge across all sectors. We clearly have to move beyond the linear conceptualization of technological progress as doing R&D to a more systemic notion that includes other actors to factor in these realities.

Scientific and technology partnerships are thus increasingly vital for latecomer countries seeking to keep up and catch up and even to move further ahead. To a certain extent, the ability to sustain large-scale in-house R&D efforts that conferred clear advantages upon larger countries and firms in the past in the agricultural sector can also be matched by the flexibility and size of the network to which smaller producers belong. These networks and partnerships provide critical support to innovative activity at the technological frontier (Wagner, 2008). Indeed, participation in such networks and the skill with which a portfolio of partnerships is managed has become essential to national ability to catch up across the board, and keep up as competitive conditions change. These are now also especially imminent to our inquiry.

Innovation underpins the cumulative increases in workers' productivity in industrial countries and is driven by technological advance, investment in physical capital, and the growth of human skills. However, these factors are shaped in very profound ways by institutions and policies. These investments in building up what we now refer to as 'systems of innovation' involves purposive actions of governments in the deliberate creation of organizations and incentive mechanisms to foster the creation, transfer, adoption, adaptation, and diffusion of knowledge. These non-market avenues are necessary (contrary to the pure market view) because as Lundvall (1988) and others suggest, the market alone is a poor filter for firm-level technical change, which is the locus of production and innovation. Other non-market coordination mechanisms are particularly important, but they are notably weak and suffer from poor systemic coordination in developing countries. Prominent among these are the structures of R&D, finance support, metrology, standards, and quality centres and, at the base of it all, the system of education, which is responsible for new knowledge from basic research and the training of scientists and engineers.[5]

The objective of innovation policies is primarily to encourage linkages between the different actors of the innovation system. This requires an integrative and holistic approach to policy formulation and demands close interaction between the different ministries whose policies have an impact on innovation and performance of the economy: education, ICT, finance, agriculture, and enterprise. In addition to the misconception that innovation is purely applicable to rich countries and the limited understanding of what innovation means in a developing context, the huge scope for formulating an innovation policy ideally designed for developing countries undoubtedly constitutes a major challenge in Africa.

1.5   Innovation capacity: Institutional basis of variations across countries

The varied institutional capacity among countries that fall under 'latecomer' points to the need for a more nuanced understanding of the processes of change and innovation and also calls for a border reconceptualization of the development process through a new technology such as biotechnology. There are three such concerns that lie at the heart of our conceptualization of the latecomer problem.

First, as opposed to the common practice of measuring innovation in more advanced countries using R&D data, which has routinely been applied to latecomer development, structural transformation that leads to the building up of R&D capabilities in latecomers derives largely from non-R&D activities. These take place at the firm and farm levels and include prototyping, reverse-engineering, incremental product, and process design, among others. Continuous interactions between non-public and public-sector organizations, where most of the R&D activities takes place in these countries, plays a critical role in such dynamism and technological upgrading.

Second, institutional and technological differences manifest themselves as costs of acquiring information and applying it to building *innovation capacity* through the market mechanism. As opposed to the dominant view in neoclassical economics, state intervention is a critical aspect of mitigating the costs through both market and non-market interventions for the generation and knowledge and learning activities. The state thus plays a major role in building productive capacity of sectors in latecomer countries.

Third, innovation in latecomer development shows distinct features. Despite a common tendency to downplay innovation in the context of latecomer development because it is not at the 'frontier', innovation proceeds in a seemingly dynamic sense in most latecomer countries. In a latecomer context, innovation is contextual and occurs despite a scarcity of resources, shaped mainly by factors such as local demand and collaboration variables among others. It therefore cannot be captured appropriately by established benchmarks such as R&D investments, patents, venture capital, and productive assets. However, the dynamism is relative, and is related to the level of institutions for physical and knowledge infrastructure as well as the state of innovation policy in the country.

It is these and other differences between latecomers and frontier countries that call attention to provide the *innovation capacity framework*, which relates non-linear dimensions of how innovation, knowledge and development interact and proceed. In a separate body of work[6], we employ a dynamic framework to analyse performance of countries over time, with emphasis on late development and factors that drive wealth creation in what we classify as 'frontier', 'fast follower', and latecomer countries. Given the countries examined in this book, we adopt the terminologies of 'fast followers' 'latecomers', and 'very latecomers' as appropriate nomenclature for these countries given their institutional capabilities for knowledge accumulation and use.

In the case of agriculture, the countries that have achieved the Green Revolution can be distinguished through the following:

(a) *Development of scientific and technological infrastructure* capacity to conduct agricultural research geared towards local needs in the national public-sector institutes and universities;
(b) *Fostering innovation policies* that focused on agricultural production as a priority;
(c) *Closing yield gaps as a result of (a) and (b)*, and
(d) *Development of physical infrastructure* to generate a comprehensive delivery system from the farmer to the market.

The 'Gene Revolution' builds further upon the innovation bases that caused the Green Revolution to succeed among countries and is configured by three main features that are path-dependent on previous investments by countries to achieve the Green Revolution, namely:

(a) *Knowledge capital* for biotechnology that build further on traditional disciplines of relevance to agricultural research;
(b) *Networking possibilities* to form hybrid organizational structures linking science to product commercialization that rely on the existence of physical infrastructure and extension services;
(c) *Regulatory and policy capacity* of countries to provide for institutions and instruments, at both the national and sectoral levels to enable biotechnology's integration into relevant activities.

We use these pre-conditions of the green revolution and the gene revolution as the basis to define latecomer countries with different levels of innovation capacity. *Fast followers* are those countries which have not only managed to achieve the Green Revolution by realizing the four parameters listed here, but have also put in place substantial investments in biotechnology-based agricultural research and,

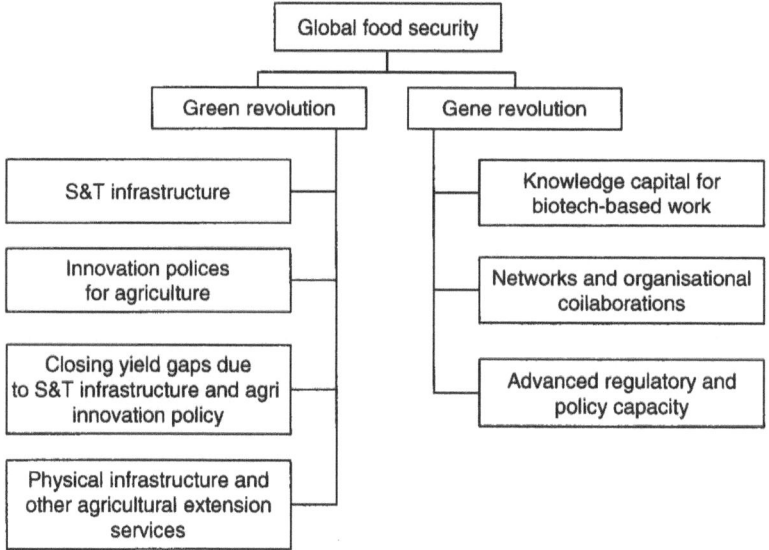

Figure 1.2 Characteristics of the Green and 'Gene' Revolutions
Source: Authors.

by implication, are succeeding in translating biotechnology research from the laboratory to the market. They comprise mostly Southeast Asian countries, and a few Latin American countries. In comparison, the *latecomers* are those countries that have succeeded in realizing the benefits of the Green Revolution but are lagging behind in bio-technology investments and capacity, and have still some way to go in enabling a Gene Revolution that could be an able successor to their Green Revolutions, in a locally relevant form. *Very latecomers* are those that have bypassed the Green Revolution the first time around and also offer a poor institutional basis for biotechnogy-based developments in their local contexts, figure 1.3.

From the foregoing, we infer that the evolutionary process of building innovation capacity will involve constant design and re-design of social and scientific organizations fitted into institutions to differing and changing contexts, while embedding local and external components within functionally coherent structures. By simply imitating structures that work elsewhere and in the past somewhere, a number of latecomer countries ignored the changing character of institutions and the organic nature of the processes that gave rise to them. This is the hard part of institutions, the continuous changes and the adaptations required that calls forth the ingenuity of the policy process.

Very latecomers
• Bypassed the green revolution the first time
Poor institutional setting for biotech-based developments in their local contexts

Latecomers
• Achieved the green revolution
• Lagging behind in biotech investments
• Still have a way to go to enable the gene revolution

Fast followers
• Achieved the green revolution
• Substantial investments in biotech research
• Translation of R&D into marketable innovation

Figure 1.3   Latecomer capabilities for agricultural biotechnology
Source: Authors.

1.6 Innovation capacity: Nature and definition

Clearly, the observed differences in national economic performance can be traced in large part to differences in their institutions (North, 1996).[7] The comparative work of Nelson (1993) on national systems of innovation showed that countries have developed different knowledge bases in both R&D and the capacity for innovation. For instance, Nelson noted the differences that size makes in systems of innovation: 'The differences in the innovation systems reflect differences in economic and political circumstances and priorities while size and the degree of influence matter a lot' (Nelson, 1993, p. 507). By implication, Nelson's definition acknowledges the role of forces outside the domain of R&D and the institutions associated with it.

Following from our discussion so far, we define *innovation capacity* as the resources required for building a complex multidimensional and dynamic range of knowledge, skills, actors, institutions, and policies within specific political policy structures to transform knowledge into useful processes, products, and services.[8] But what elements make up innovation capacity and what differentiates them across latecomers? There are four broad dimensions that help to answer this question.

First, the quantum of inventive activities including but not limited to R&D[9] (an important source of learning for innovation) carried out in universities and firms are significantly lower than is found in advanced industrial countries and equally vary between latecomers and very latecomers. In latecomers much of the innovative activities in firms are imitative and product-related rather than process-centred. Again, the functions of the production systems are different. For instance, industrial production in the US is more specialized in R&D-intensive hi-tech products – far more so than in the European Union (EU); and public-sector research, for example at universities, is more closely linked to industry, performing R&D functions that private-sector firms fulfil in Japan, for instance (Edquist and Texier, 1996).

Second, the competence building capacity of organizations such as universities and training centres, many of which were set up expressly to produce manpower, is relatively small and in the very latecomers has failed to meet the challenges of the new, more competitive global economy (Oyelaran-Oyeyinka and Barclay, 2003).

Third, the function of information exchange is usually very weakly coordinated or not coordinated at all in the very latecomers. In latecomers, the situation is better, although often still imperfect. When compared to the systems of innovation in advanced economies, the flow of information is much greater and access to it is generally easier, even for non-specialists, although a significant proportion of R&D information is withheld from the public domain because it consists of trade secrets.

Fourth, the capacity for regulatory functions and enforcements of systems of innovation differ, even among advanced countries, but these differences are more pronounced still between latecomers and some have almost no regulatory institutions for dealing with imported new technologies. These are further called into question by international rules and external pressures that call for delicate balancing between intellectual property rights obligations, biosafety regimes, and local development needs. Countries need to make informed decisions on environmental risks associated with agricultural biotechnology not only to devise optimal biosafety regimes, but also to ensure that they do not shun away the potential benefits that agricultural biotechnology holds due to premature environmental concerns. Second, the international trade regime as created under the World Trade Organization (WTO) and related agreements already prescribe intellectual property rights over plant varieties (in the Agreement on Trade Related Aspects of Intellectual Property Rights, 1995 – the TRIPS Agreement) and their terms of trade. Therefore, how countries go about regulating environmental impact of GM technologies depends largely on how and to what extent they are able to create a balance between rights and obligations in the current intellectual property regime on agricultural biotechnology to minimize environmental damage (Tansey, 2002, p. 22). Thus, meaningful environmental/biosafety regulation of GM technologies requires sorting out of the interface between the TRIPS Agreement that provides for intellectual property rights on plant varieties and the Cartagena Protocol on Biosafety.

Hosts of *very latecomers* are at very early stages of developing biosafety systems to regulate the introduction and release of GM organisms. Most of them have adopted *sui generis* regimes for plant variety protection under article 27(3)(*b*) of the TRIPS Agreement that may not suit their local development needs. Very importantly, the skewed focus on genetically modified crops and the precautionary principle (see Box 1.3) in latecomers and very latecomer countries could in fact be a large reason for the lack of policy vision in to explore the potential benefits of agricultural biotechnology that span so much beyond GM technologies.

Box 1.3 Plant variety protection and biosafety regimes in latecomers

The TRIPS Agreement and the Cartegena Protocol on Biosafety are two important international agreements that in recent times have sought to set standards for agricultural biotechnology products and related trade.

Article 27(3)(*b*) provides for the protection of plant varieties either through patents or an effective *sui generis* system (a system of its own). Pursuant to the TRIPS Agreement, the 1991 Agreement of Union of Plant Varieties (UPOV) has been advocated as an appropriate *sui generis* regime for latecomer countries and many of them have also adopted it. There are several reasons why this may not be the best option for farmers and agricultural innovation systems in latecomers. These are discussed in Chapter 2.

The Cartagena Protocol on Biosafety that came into force in September 2003, is the main international agreement that provides guidelines to national governments as to how they can best address issues of biosafety within their countries for unknown risks to human health and the environment that remain unproven by almost two decades of biosafety research (Juma, 2001). The debate on risks to the environment and human health due to GM technologies is ongoing and involves questions of a different nature such as: How should countries implement the 'precautionary principle'?[10] How important is labelling of GM products to ensure food safety and consumer protection? How much and how far can uniform guidelines solve such issues. If international guidelines are indeed the answer, can the Codex Alimentarius (a compilation of food standards worldwide) be such an international option?

Source: Authors.

In sum, the innovation capacity framework, with its roots in evolutionary economics exhibits considerable heterogeneity and it is the differences in these institutions that provide four key factors (interactive learning, knowledge bases for biotechnology, incentives for translating inventive activities, and state policy capacity) for defining and comparing *innovation capacity*. We use the sectoral systems of innovation framework that is primarily a sub-discipline of innovation studies, aimed at understanding the underlying factors that contribute to variances in sectoral performance and their contributions to economic growth of countries, in order to compare innovation capacity of latecomer countries in agricultural biotechnology.

1.7   Innovation policies and innovation capacity

The categorization of countries at three levels, namely fast followers, latecomers, and very latecomers suggest that policies are differentiated by differences in levels of development. The role of policies in strengthening learning, investment and linkages that constitutes the bases for dynamic innovative change on a continuous basis is therefore critical. There are a variety of policies including policies affecting size and shape (demand characteristics) of the domestic market, for example, taxation, wages; policies that affect input costs or outputs for entrepreneurs, for example, land prices and use; policies that change the nature of competition, foreign investment, and those that promote local upgrading and linkages between foreign and local agents; policies that change or make possible access to training for vendors and manufacturers, and international rules that affect learning and innovation.

Innovation policies and science and technology policy consists of both horizontal and targeted policies. Horizontal policies are directed to stimulate technological development irrespective of specific technological area or industry (Teubal, 1998). In the case of agricultural biotechnology, science, technology, and innovation policy will include both horizontal and targeted policies such as:

- R&D investment (government funding of research);
- Government technology procurement: This is a relatively common policy tool used in industrialized countries such as Japan and the United States (even though their models of procurement differ[11]). Procurement constitutes an important mechanism for the articulation of demand, and some authors think it constitutes a special case of user–producer interaction (Edquist and Hommen, 1998b). Apart from considering the implementation of the government technology procurement tool, it is necessary to analyse whether or not the procedures behind the procurement regime reinforce open competition among firms. It could be the case that the procurement regime has a reverse effect, favoring a specific type of firm. Government procurement could be complemented with other technological policies such as R&D subsidies or reinforcing the articulation of demand;

- Standard setting: The setting of standards often requires the public and private sectors to interact. The government might also set standards through government procurement. In any case, the setting of 'good' standards is essential for enhancing innovation. Higher standards are recommended given their role for competitiveness (Edquist and Hommen, 1998b);
- Training and education: Not only does the availability of highly educated and trained skilled workers matter, but also the orientation of training and education programmes towards problem-solving, and their linkages with industry;
- Collaboration-oriented policies: collaboration is central to interactive learning and policies for collaboration can take different forms depending on actor configurations. Prominently, they include interactions such as:
  - Supplier–producer (university–industry programmes)
  - User–producer (agricultural producers–biotechnology firms for articulating demand)
  - Producer–producer (biotechnology firms)
  - User–producer–supplier [through interface organizations (e.g., research technology organizations) making a bridge between science-based organizations and industrial practices]
  - Support organizations for business firms (including government agencies).

The policy competence to enact and match different policy instruments (both horizontal and targeted) has been instrumental in the economic growth trajectories of today's industrialized countries and also accounts for sectoral performance among sectors that employ new technologies.

It is precisely these differences that brings us to the central questions that inform the analysis in this book. In a latecomer context, how are firms and scientific organizations responding to the advent of biotechnology for agriculture? What are the factors that determine policy vision of countries towards agricultural biotechnology, and the ways and means in which states perceive its role to solve the most fundamental issue in development: food security? What institutional forms of collaboration have emerged and how can these be understood? What is the nature and specifics of the collaboration and how intense are they? Where are the gaps and what are the obstacles to collaboration given that no single organization can undertake the kinds of activities necessary to produce high-yielding, biotech-based crops in latecomer countries?[12] Are there incidences of organizational learning through collaboration, where we regard learning as socially embedded and highly conditioned by the community in which science and production is practiced? As Brown and Duguid (1991: 48) have perceptively stated 'learning is about becoming a practitioner, not learning about a practice'; in other words is a community of practitioners of the new technology emerging in latecomer countries? Literature tends to suggest that the institutions of networking, as opposed to the notion of a single firm vertically integrating from the laboratory to the market has become the dominant form of operation in biotechnology, and as such, the extent of a firm or organizational reach into these networks determines the capability of the organization. Do these observations hold true in latecomer countries and if not, what innovation capacity constraints help explain their absence? And, most importantly, what is the role of policies and insttutions in promoting agricultural biotechnology systems of innvoation, and related Gene Revolutions of relevance to their local contexts and needs? What insights does a comparative analysis of different countries and their innovative capacity constraints, as undertaken in this book, provide towards designing successful sectoral systems in agricultural biotechnology that can address poverty and food security issues in latecomers? In conducting the analysis, we are specifically concerned with how the so-called 'orphan' crops such as sorghum, millet, cassava, yams, sweet potato that are not bait enough for big seed corporations can be developed through national or regional initiatives between latecomers themselves.

1.8   Methodology

The research informing this book was motivated by two factors: first, an innovation system-oriented and policy-relevant innovation survey for biotechnology is complex and not too many such surveys have been conducted in latecomer countries. Few innovation surveys carried out in the 1990s, for example, were consciously designed for policy relevance. The focus in most of these, was less on learning and innovation processes and more on innovation inputs and outputs. But sample frames were easier to create and there was a depth of knowledge about sampling and surveying that came from the experience acquired in carrying out both industrial and science and technology surveys. Second, the debate on capacity-building in biotechnology has proceeded with a casual disregard of existing capabilities in latecomer countries which may be used to create successful sectoral systems of biotechnological innovation. Therefore, this research considered as its focal points, the need to enhance policy relevance, sharpen the focus on core elements in an innovation process – linkages, learning, and investment and to ensure the quality of data and the manageability of the survey process.

Our research followed a three stage process: background reports, questionnaire surveys, and interviewing. As a first stage, background reports on the state of agricultural biotechnology were created in collabroation with national organizations in all the six countries that used governmental data and secondary sources. Using background information generated by the different national collaborating organizations, actors were identified within the agribiotech system of innovation, and the critical interactions between them were mapped. The following category of actors were considered for the survey and case studies: public and international research institutions; university departments such as microbiology, botany, agricultural sciences, genetic engineering; private firms involved in biotechnology (both local and foreign); international research institutions; as well as government departments, and agencies involved in certification and regulation.

Semi-structured questionnaires designed to capture the nuances of the innovation environment were administred to each of these groups of actors in all the six countries, and these were followed by interviews conducted (face-to-face) with high-level executives from firms, and groups of scientists as well as directors and CEOs of governmental agencies. On the whole, 264 firms and public-sector organizations were surveyed in Vietnam, 210 firms and public-sector organizations were surveyed in Nigeria, 76 firms were surveyed in Malaysia, and 74 firms and public-sector organizations were interveiwed in Kenya.

We define innovation as 'a product, process or organizational change that "new" to the firm/organization even if the innovation is not new to the country or the world at large'. We have adopted this definition for two reasons. First, agricultural biotechnology as a sector represents learning activities on a continuum of knowledge spanning from a range of conventional to new technologies, and the capacity to build further through a step-by-step process of innovation will determine competence. Firm or organizational level learning, in such a context, occurs through small increments to the existing pool of technological capabilities, however close or distant they may be to frontier science. Second, our focus is on arriving on a set of policies that promote processes of learning and innovation in contrast to those that focus solely on quantitative goals in the production of inputs or outputs. Such a definition of innovation serves this goal better.

The research identified the triggers to innovation among these major actors, followed through some of the processes and identified the obstacles to creating dynamic sectoral systems of innovation for agricultural biotechnology. Our main emphasis was on inter-organizational interactions and their impact on performance and innovation in the agri-biotechnology sector within the countries. We employ both qualitative and quantitative approaches to understand the nature of interaction and to identify the innovation capacity of countries, which are presented in the relevant country chapters.

2
Sectoral Systems for Agricultural Biotechnology

2.1 Introduction

The shift from the previous conception of knowledge production and use as something originating exclusively out of research and development (R&D) has been gradually replaced with the conception of knowledge utilization as the building of a national and sectoral innovation system. In this book, we adopt the concept of *innovation capacity* to describe the knowledge as well as the institutional structures in order to use scientific and technological knowledge for commercial products and processes. In other words, innovation capacity lies at the root of the systematic differential performances of countries and sectors and is responsible for the economic advancement of the more industrialized countries. We differentiate between countries that are traditionally associated with frontier knowledge and the rest. This book and its analysis is not about those at the frontier of technology and innovation, it is about those latecomer countries trying to catch up with the industrial leaders. In so doing we make a distinction between *fast followers, latecomers,* and *very latecomers;* terms that are fully described in subsequent sections.

Our current understanding of the process of change in different contexts shows that, far from being a just a technical phenomenon, innovation is a social process shaped by a wide variety of factors and is understood only within specific historical and institutional settings in which the key actors are located and defined by the intensity of their interactions. This approach to knowledge and capabilities does away with the decades of debate on science and technology (S&T) promotion and development that positions supply-side factors, inspired by the linear model of science, as the main focus of policy.

Another important distinct emphasis of the book is the call to commitment in science technology and innovation (STI) investments by latecomer countries for their own immediate internal necessities of food security and not just for the opportunity to participate in knowledge-driven global markets. Latecomers need to make deliberate and purposive investment in building policy capacity to construct capabilities for managing their 'systems of innovation', the repercussions of which are so much more far-reaching than merely enhancing their competitiveness or productive capacity. In other words, latecomer governments need to make deliberate efforts to build policy capacity in addition to, as well as establishing the incentive regimes to foster the creation, transfer, adoption, adaptation, and diffusion of knowledge. This book argues for a central role of the state in building the system of innovation for agricultural biotechnology.

2.2   Applying innovation systems framework to late development

Innovation is often confused with research and measured in terms of scientific or technological outputs. Recent literature, however, stresses that innovation is neither research nor S&T, but rather the application of knowledge in production. This knowledge might be acquired through learning, research, or experience, but until it is applied in the production of goods or services it cannot be considered innovation.

The range of innovations is quite wide, comprising both radical changes and many small improvements in product design and quality, in production processes or the way in which production is organized, in management, marketing or maintenance routines that collectively, modify products and processes, bring costs down, increase efficiency, and ensure environmental sustainability. As opposed to the focus on novelty that is central to the concept of invention and a key criterion for patenting, innovation is a broader concept and consists of all products, processes, and organizational forms that are new to the firm in question, irrespective of whether they are new to their competitors, their country, or the world at large.

Innovation can be triggered in many ways. Bottlenecks in production within a firm, changes in technology, competitive conditions, international rules or domestic regulations, environmental or health crises, and even wars have been known to stimulate a process of innovation (Rosenberg, 1976; Dosi, 1988; Chandler, 1990; Nelson, 1996). Over the past several decades, a number of changes in the pattern of production and competition in the world economy have been particularly important in shaping the relationship of innovation to equity, competitiveness, and sustainable development.

During the 1970s and 1980s production became more knowledge-intensive as investments in intangibles such as R&D, software, design, engineering, training, marketing, and management began to play a greater role in the production of goods and services. Within the context of more knowledge-intensive production, firms began to compete not only on price but also on the basis of their ability to innovate. As traditional barriers to trade and investment were dismantled, innovation-based competition diffused around the globe. This put pressure on local firms everywhere to engage in a process of continuous innovation and challenged governments to develop policies to stimulate and support innovation processes.

Conventional economic models that viewed innovation as a linear process driven by the supply of R&D, however, were increasingly subjected to criticism for their limited explanatory power and lack of guidance for policymaking under these changing technological and competitive conditions. This created space for the emergence of alternative conceptualizations of the innovation process notably those that understood innovation in a more systemic, interactive, institutional, and evolutionary terms. Over time, these 'innovation systems approaches' gained wide support among OECD member countries and more recently have been applied in the European Union and in a number of developing countries as a framework for policy-relevant analysis (OECD, 1992; Wong, 2003; Cassiolato et al., 2003).

A number of features differentiate an innovation system from both the traditional production-oriented, equilibrium-based models of the economic system and from the narrower focus on S&T systems that were an earlier effort at dealing with the role of technological change in economic development. An innovation system is conceptualized as a network of firms and other economic agents who, together with the institutions and policies that influence their innovative behaviour and performance, bring new products, new processes, and new forms of organization into economic use.[1] As an evolutionary system, the focus is on interaction between these actors and their embeddedness in an institutional and policy context that influences their innovative behaviour and performance. Table 2.1 summarizes the static and dynamic conceptions of innovation and innovation capacity.

Table 2.1 Mainstream economics and evolutionary theory: Issues for characterizing innovation capacity

Innovation capacity & innovation policy issue	Mainstream economics	Evolutionary theory
1. Theoretical explanation	Equilibrium analysis	Dynamic processes of variation, selection, and reproduction.
2. Conception of knowledge	Codified information explained in terms of production function: blueprints and drawings.	Tacit knowledge is as important as codified knowledge. The package comprises both codified and tacit knowledge and there is emphasis on the use of new technology.
3. The growth of knowledge and development of innovation.	One homogenous input source: R&D	Knowledge growth through search, learning, and interaction between firms and organizations.
4. Actor and activities involved in the development of knowledge and innovation	The key actor is the advanced industrial firm and activity is the R&D function of advanced/'high technology' manufacturing firm.	Capabilities developed across firms and organizations mediated by institutions; knowledge involves both high-tech, traditional, and service sectors.
5. Transaction costs involved in the diffusion, adoption, and use of knowledge and innovations and in transfer processes.	The model assumes zero transfer and absorption costs.	Explicit investments in resources (financial, human knowledge, and skills) to adopt, absorb, use, and validate both codified and tacit knowledge. Firms and countries require 'absorptive capacity' and continuous interactive learning within the right institutional context and policy support.
6. Concept of collaborative learning	Not a required condition	Collaborative learning is a basic premise of systemic innovation with wide diversity in the structural characteristics of firms. While some firms possess the knowledge others do not have adequate resources to sustain continuous learning and innovation.

(Continued)

Table 2.1 (Continued)

Innovation capacity & innovation policy issue	Mainstream economics	Evolutionary theory
7. Variety and diversity (firms, sectors, and organizations)	No scope for organizational and firm diversity; actors are homogenous.	Fundamental heterogeneity in the structure of firms and organizations and as such considerable differences in the learning and innovation capacity at firm, sectoral, and national levels.
8. Selection mechanism	Purely market, non-institutional (exception is a focus on intellectual property).	Market and non-market institutional mechanisms.
9. Innovation process	History does not matter	Path-dependent, cumulative processes with multiple feedbacks.

Source: Oyelaran-Oyeyinka (2006).

The scope of potentially important economic actors in an innovation system also differs from the set of suppliers and clients arrayed along the classic value chain and incorporated into input-output models or from the set of organizations – universities, public-sector research bodies, and science councils – that are the traditional focus of S&T studies. There is no assumption, moreover, that an innovation process is linear or that knowledge outputs feed directly or automatically into products for sale in the market. Instead, the knowledge and information flows that are at the core of an innovation system are multidirectional in nature and open opportunities for the development of feedback loops that can enhance competence-building, learning, and adaptation.

2.2.1 Innovation systems and late development

All too often in the case of latecomer countries, however, the right kind of actors are not present or do not interact in a way that stimulates or supports a process of innovation. The innovation system approach provides a framework that is useful in understanding why this is the case. Three of its features are particularly important in this respect.

First, a distinction is drawn between 'organizations' such as enterprises, government ministries, non-governmental organizations (NGOs), professional associations, R&D institutes, innovation and productivity centres, extension services, standard-setting bodies, universities and vocational training centres, information gathering and analysis services and banking, and other financing mechanisms as actors in an innovation process and 'institutions'. The latter, are understood as 'sets of common habits, routines, established practices, rules or laws that regulate the relations and interactions between individuals and groups' (Edquist, 1997, p. 7) that 'prescribe behavioural roles, constrain activity and shape expectations'.(Storper, 1998, p. 24). Simply co-locating potentially critical actors co-located within a geographical space is a necessary but not a sufficient precondition to interactive learning. The nature and extent of interactions are rather structured through the institutional framework that provides both tangible and intangible incentives for actors to perceive mutual gains from collabroation.

Second, the innovation system approach factors in the demand-side of innovations, thus centring attention on local demand for particular products/processes, such as those for particular crops, medicines, or essential goods that form part of development concerns. Demand flows are among the signals that shape the focus of research, the decision as to which technologies from among the range of the possible will be developed and the speed of diffusion of these technologies. Demand is not solely articulated at an arm's length through the market, but may take place through a variety of non-market mediated collaborative relationships between individual users and producers of innovation (Lundvall, 1988, p. 35). In still broader terms demand may be intermediated by policies. Enhancing knowledge and information flows is yet another way to stimulate innovation and facilitate adaptive policymaking.

Third, the innovation system approach acknowledges that policies matter. Whether tacit or explicit, policies play a role in setting the parameters within which actors make decisions about learning, investment, and innovation. It also recognizes that policy dynamics supportive of an innovation process are not the outcome of a single policy but of a set of policies that collectively shape the behaviour of actors. The need for an overall innovation strategy, for priority setting and for policy coordination is thus critical in strengthening innovation systems whether at the national, local, or sector levels. From a policy perspective, the innovation system approach has a number of important strengths. It builds awareness, for example, of the extent to which norms, rules, and institutions are learned behaviour patterns marked by the historical specificities of a particular system and moment in time. As such, their relevance may diminish as conditions change. Policy dynamics, moreover, are generated by the interaction of policies with the behavioural norms and attitudes of actors that they seek to condition. Learning and unlearning on the part of all actors including policymakers are thus essential to the evolution of an innovation system in response to new challenges.

Monitoring the policy dynamics generated by the interaction between policies and the varied norms and rules of actors in the system and opening channels for dialogue, for example, would be of importance in fine-tuning policies for maximum impact and responsiveness to changing technological and competitive conditions. Policies thus have an important role to play in reinforcing older norms and rules or in stimulating and supporting change.

By refocusing on the process of innovation as opposed to quantitative indicators of inputs or outputs, the innovation system approach encourages the adoption of a broader perspective on the set of actors that might be critical to innovation in a specific sector or within the broader national or local economy. It also induces the adoption of a longer-term perspective on the recombinatorial and reuse opportunities that lie in networking as a means to strengthen the capabilities and widen the variety of local knowledge bases. Its dynamic strengths are also evident in the stimulus it provides for a reconceptualization of sectors as potential 'innovation systems'. Where enzyme technologies are developed as an input to the mining sector, for example, a longer-term and larger set of linkages might be envisaged. This provides a more dynamic way of dealing with the uncertainties resulting from changing technological, economic, and competitive conditions.

While it is not possible to predict, *a priori*, the direction of technological change and thus identify the specific actors that might be critical to the innovation process in the future, networking creates a more flexible structure for responding to change. This is particularly important today when the institutional set-up at the global level has become a powerful force that shapes the parameters within which actors make critical decisions with respect to innovation.

2.2.2 Sectoral innovation systems to understand agricultural biotechnology

While knowledge systems are universal – cutting across economies and industries – sector specifics often demand the development and diffusion of specialized knowledge. A sectoral innovation system has its own knowledge base and learning processes, it has specific technologies, systems boundaries, firms, institutions, and interactive activities, and yet it is connected indirectly with institutions, firms, and other systemic elements outside its own sector boundaries. Some institutions and firms connect wholly with particular sectors, while others interact directly with two or more sectors. Some institutions (e.g. basic service and utility organizations such as security, customs, water and electricity boards, and high-tech organizations such as universities) connect with all sectors. The sectoral system of innovation allows the capture of all elements that firms – as the most dynamic elements – connect with for driving learning and innovation directly or indirectly, consciously or unconsciously, and destructively or constructively.[2]

The argument for the application of the sectoral system of innovation framework is that it is able to address the evolutionary problem of capacities development that assist firms, sectors, and countries in coping with and competing in dynamic, ever-changing technical, economic, and institutional environments. Furthermore, the capacity to innovate will involve a system of diverse organizations or actors, notably the private sector but also others outside of the state, whose actions are shaped by a variety of institutional, policy, market, and technological signals. The framework is therefore particularly suited to exploring emerging, dynamic sectors where the private-sector and other non-governmental actors are playing leading roles and where firms, sectors, and countries have to cope with shocks and deal with competitive pressures.

Across sectors and time, moreover, different configurations of critical actor will emerge from among the multitude of firms and other organizations – industry associations, R&D and productivity centres, universities, vocational training institutes, information gathering and analysis services, engineering services, banking and other financial mechanisms, standard-setting bodies – whose relationship to the innovation process at a sector or system-wide level might prove critical, yet today we have little information about the range of actors that currently exist in the local/national or sector context, their competences, habits, and practices of learning and interacting or the propensity to innovate. How different social norms, practices, and other institutions affect the processes of learning and innovation in a given national or regional context is also poorly understood. Policies are rarely monitored or evaluated thus limiting our ability to assess the way in which current policies affect the parameters within which the decisions of local actors with regard to learning, linkages, and investments take place.

In applying the sectoral innovation system framework to agricultural biotechnology in latecomer countries, this book directly challenges two of the beliefs still common in development circles. The first is the traditional approach to building local technological and productive capabilities in developing countries. It regards developing countries as 'technology users' and emphasizes the simplistic North–South technology transfer. From this perspective, learning is treated as largely imitative as opposed to innovative while catching up is an incremental process from a low-wage, low-skill base in which tertiary education and local research capacity play only a minor role at best.

The belief that latecomer countries are users rather than producers of technology has also contributed to the slow emergence of policies appropriate to current development needs and competitive conditions. From this earlier perspective, catching up was understood primarily as a process of technology transfer from leaders to latecomers and the absorption of imported technology was believed to be automatic and costless. Conventional theory envisaged latecomers largely as passive recipients of technology from elsewhere (Katz, 1985, p. 127), thereby missing out on the importance of tacit know-how and local knowledge for technology absorption and use. The benefits of technology transfer were simply assumed to 'trickle down' and 'diffuse' without impediments. Recent developments in the application of biotechnology to agriculture (and those related to pharmaceuticals and health)[3] and the debates over the digital divide have reinforced the fundamental misconception in conventional logic that importing foreign technology and creating it locally are not alternatives but complements (Bell & Pavitt, 1997).

The second is the notion that innovation is something that takes place only in countries such as Japan or the US, in large multinational corporations, or in what are regarded as the high-tech industries. Indeed, much conventional literature continues to associate innovation with the kind of activity undertaken by firms that takes place at the technological frontier. A narrow definition that equates innovation with invention of this sort denies the importance of:

(a) Building upon existing indigenous knowledge,
(b) Exercising creativity in the development of new products, processes, management routines or organizational structures that correspond to local conditions and needs,
(c) Creating the local linkages that support the modification of production processes to bring costs down, increase efficiency and ensure environmental sustainability,
(d) Mastering imported technology in order to transform it in new ways and
(e) Developing policies that stimulate and support a continuous process of learning and innovation.
(f) Local demand.

Viewed on a continuum of competitiveness, innovation implies continuous improvement in product design and quality, changes in organization and management routines, creativity in marketing, and modifications to production processes that bring costs down, increase efficiency, and ensure environmental sustainability. The ability to manage complex partnerships and linkages and to learn through them, overcome local constraints, and harness knowledge-based tools are firm and organization-specific advantages in innovation-based environments.

This is no more true than in the present knowledge-driven context of global activities wherein sectoral systems contribute to the domestic economy and also enhance international competiteveness of national activities by forging success. A sectoral systems perspective shifts the focus onto:

(a) Opportunities for learning and innovation in small and medim enterprises (SMEs) and in the so-called 'traditional industries' much more than they have done in the past.
(b) Need for knowledge-based resources to enable learning at the sectoral level that create channels for knowledge flows and provide the stimuli and support needed by potential users to overcome the uncertainties, costs, and risks associated with a process of innovation.
(c) The innovation capacity constraints posed by the institutional embeddedness of the larger national system of innovation that impinges upon the ability of the sector to learn and grow dynamically.

2.3   Agricultural biotechnology as a sectoral system of innovation

In its most basic form, a system has actors, boundaries, and interactions to promote learning and access to knowledge sources that in turn build the innovation capabilities of actors. A sectoral system of innovation analysis seeks primarily to ascertain the boundaries of the system. In other words, what are the peculiar features of the system at the sectoral level that demand immediate and specific attention, although several attributes remain generic at the national level?

Agricultural biotechnology as a sectoral system of innovation has certain attributes that fall back on the technological characteristics, on the one hand, and the dominant industrial trends in frontier countries, on the other. For one, biotechnological tools that form the basis of the Gene Revolution, although largely a result of academic research, rely heavily on their uptake by established seed and chemical companies for product development. Seed and chemicals, as a global sector, is oligopolistic in nature, where innovation once again is highly intellectual property-oriented, and dictated by the needs of the developed countries. As noted repeatedly in the literature, the majority of the private-sector investment in agricultural biotechnology is indeed focused on four crops – cotton, canola, corn, and soybean (see Pingali and Raney, 2005; STEPS, 2008). The private-sector bias thus is a bleak prospect for crop development that focuses on ensuring food security in developing countries.

Biotechnology growth in the OECD countries has been promoted mainly through government-based R&D, which figures as the most important policy instrument for the sector, followed by other regulatory measures that allow for commercialization of research results, technology transfer, and the creation of intermediary organizations. Recent research on the policies that promoted biotechnology in 15 EU states and three associated countries found that between 2002 and 2005, government-funded R&D was the primary policy instrument with biotech R&D funding as part of total government R&D spending seeing a rise in the majority of the 18 countries surveyed. The survey also found that at least 201 instruments were in place in the countries to support commercialization of products of which 71 were specifically biotechnology-oriented (Enzing et al., 2008).

Finally, regulation is a major element in biotechnology, due to its science-intensive nature and inherent perils that are sought to be regulated through biosafety rules and regulations. Policy frameworks need to include policies that provide a critical science base, collaboration incentives, risk minimization associated with commercialization of products, extension policies that eliminate information asymmetries among farmers, promote uptake of new locally useful varieties, and enhance farm-level production capacity and access to markets.

Accepting these to be the dominant features of systems of innovation in agricultural biotechnology, we are forced to ask the obvious question: Do agricultural biotechnological systems exist at all in latecomer countries where the private sector is largely absent, governmental R&D spending is marginal, and technological capacity is not well advanced? If we were to assume so, what are the points of departure of agricultural biotechnology systems of innovation in late development?

Three major factors stand out when analysing agricultural bio-technology in latecomer countries. The first attribute of agricultural biotechnology in a latecomer context is that the private sector is conspicuous largely by its absence, rather than for its proven ability for product development as is the case in the frontier countries. The systems of innovation at the sectoral level are quite often so stymied by the absence of private enterprise that most research results from the public sector do not find their way to the market. This phenomenon, which we have in the past termed 'the rough road to the market', is a commonplace occurrence in latecomer countries across all sectors (Gehl Sampath and Oyelaran-Oyeyinka, 2009). The point to be highlighted here is two fold: the knowledge base is more dispersed than what we know from our experiences of studying agricultural biotechnology in the frontier countries, and the organizations that play the critical role in applying existing knowledge or generating new biotechnology knowledge through learning activities are in the public sector.

Second, the market for agricultural products is severely fragmented in latecomer countries and this stunts advances that require demand- and supply-side coordination. In other words, local demand never gets codified into local research or innovation agendas due to information asymmetries within user–producer networks (Stiglitz, 1982) as a result of which directed and targeted investment into R&D capacity (even in the public sector) does not materialize. The system of market exchange in which productive agents are embedded is overwhelmingly imper-sonal and often subject to high levels of uncertainty due to the equally uncertain nature of relational contract.[4] Observing the same in the case of machine-tools sector in Taiwan, Amsden (1977, 1985) highlights the ubiquitous ways in which the *size* and *type* of markets shape the rate of knowledge creation as well as the division of labour. The 'extent of market' or 'size of market' refers not to a geographic area or large popu-lation but rather to purchasing power, 'the capacity to absorb a large annual output of goods'.[5] She makes a distinction between the notion of 'size' and 'type' of market, which applied very well to agricultural markets in latecomers. There may be two markets of equal purchasing power but qualitatively different in their capacities to consume large amounts of goods.[6] Markets in latecomer countries have relatively small size (thrive on personal exchanges of kinship relations, personal loyalty, and social connections)[7] and fit in very many respects with types of markets that are characterized by low profitability, limited economies of scale, and low-intensity learning that slows long-run technological capability-building. The emergence of markets has an innate structural relationship with the institutions that govern a particular sectoral system/economy or country. As North (2005, p. 123) states:

> The performance characteristics of each market will be a consequence of both formal rules and informal norms of behaviors that modify, qualify or even negate formal rules. The transaction costs in each market will reflect the combination of formal and informal constraints.

Again, in addition to articulation of demand, the technologies that underlie the Gene Revolution (such as transgenic technologies) are different from Green Revolution technologies in the sense that they are transferred primarily through markets (Pingali and Raney, 2005). This once again calls attention to the information asymmetry issues that the state needs to address.

Finally and, perhaps most importantly, technological advances in biotechnology have proceeded in an uncoordinated way and are quite unsystematic and largely at its infancy in latecomer countries. As of 2004, for example, China was the only developing country with transgenic crop technologies and India and Brazil were in the process of developing the same (Raney, 2004). The cumulative aspect of technological advance requires that countries need to have built *a priori* capacity for traditional agricultural research to advance to the new Gene Revolution. The historical constraints that have prevented a large set of countries from enjoying the benefits of the Green Revolution remains a major impediment to the emergence of capacity in agricultural biotechnology. In the case of agriculture, the countries that have achieved the Green Revolution can be distinguished through the following:

(a) *Development of scientific and technological infrastructure* capacity to conduct agricultural research geared towards local needs in the national public sector institutes and universities;
(b) *Fostering innovation policies* that focused on agricultural production as a priority;
(c) *Closing yield gaps as a result of (a) and (b)*; and
(d) *Development of physical infrastructure* to generate a comprehensive delivery system from the farmer to the market.

Success in agricultural biotechnology, the pathway to the 'Gene Revolution', builds further upon the innovation bases that caused the Green Revolution to succeed among countries. It is configured by three main features that are path-dependent on previous investments by countries to achieve the Green Revolution, namely:

(a) *Knowledge capital* for biotechnology that build further on traditional disciplines of relevance to agricultural research.
(b) *Networking structures* to form hybrid organizational structures linking science to product commercialization that rely on the existence of physical infrastructure and extension services.
(c) *Regulatory and policy capacity* of countries to provide for institutions and instruments, at both the national and sectoral levels to enable biotechnology's integration into relevant activities.

The range of knowledge activities that capture agricultural biotechnology of relevance to latecomer development are (OECD, 2002, p. 40):

(a) DNA-based, including genomics, pharmacogenetics, gene probes, DNA sequencing/ synthesis/amplification, and genetic engineering and genetic modification techniques.
(b) Proteins and molecules-based techniques, including protein sequencing, proteomics, hormones and growth factors research, and cell receptors/signalling/pheromones.
(c) Cell and tissue culture and engineering, including tissue engineering, marker-assisted selections, cellular fusion, hybridization, embryo manipulation, and vaccine/immune stimulants.
(d) Process biotechnology, including bioreactors, fermentation, and bioprocessing.

However, many countries have made no progress towards putting these prerequisites for a Green Revolution in place and therefore their ability to create and sustain agricultural biotechnology systems of innovation that rely on these three features is under constant intense scrutiny.

We use these preconditions for the Green Revolution and the Gene Revolution as the basis to define latecomer countries with different levels of innovation capacity. In the framework for agricultural biotechnology developed in the remaining sections of this chapter and applied throughout the book, we distinguish between fast followers, latecomers, and very latecomers on the basis of their Green Revolution achievements, on the one hand, and their ability to create the innovation capacity for biotechnology-based progress (the Gene Revolution that builds upon the Green Revolution success), on the other. Figure 2.1 depicts the trajectories of capability formation in latecomers in this regard.

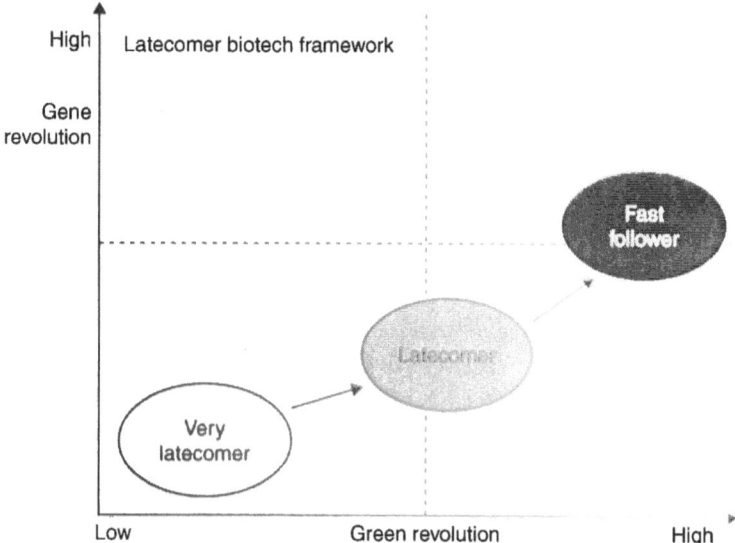

Figure 2.1   Capabilities formation for agricultural biotechnology within latecomers
Source: Authors.

What differentiates fast-follower countries from latecomers and very latecomers is the degree to which they have succeeded in achieving not only the Green Revolution, but have also managed to build the appropriate institutional and knowledge bases for a agricultural biotechnology-based Gene Revolution geared to their own local needs. Fast followers have not only managed to achieve the Green Revolution by realising the four parameters listed here, but have also put in place substantial investments in biotechnology-based agricultural research and, by implication, are succeeding in translating biotechnology research from the laboratory to the market. The latecomers are those countries that have succeeded in realizing the benefits of the Green Revolution but are lagging behind in biotechnology investments and capacity and have still some way to go in enabling a Gene Revolution that could be an able successor to their Green Revolutions in a locally relevant form. Very latecomers are those countries that have been bypassed by the Green Revolution the first time around and also have a poor institutional basis for biotechnology-based developments in their local contexts. Those countries that did not manage to develop the capacity to exploit the Green Revolution in the first place will find it harder to set up agricultural biotechnology systems to achieve the Gene Revolution simply because of the path-dependency character of technology-led economic development as well as the convergent nature of knowledge bases. While the analysis in this book explores mainly the innovation capacity of countries for agricultural biotechnology, the issue of path-dependency between the Green and Gene Revolutions is of critical importance in understanding agricultural technologies and food security in latecomer countries. This is examined in depth in Chapter 7 (comparative insights).

2.3.1 Main actors and networks

Actors or agents operating in the sectoral system include individuals such as farmers, enterprise owners, and engineers/scientists; and organizations including enterprises, universities, and firms, R&D departments, financial institutions such as development banks, and intermediary organizations such as seed banks. There are important attributes of the actors that mediate the innovation process including ownership structures (whether firms are owned by multinationals or local entrepreneurs), size and extent of local enterprise, quality of local research institutes, among others. Ownership structures create different sorts of incentives with regard to innovation. The size of enterprise is equally important for its relationship to choices of product, techniques of production, its ability to generate capital for investment, and is usually also related to the level of education of owners and the opportunities available to them for learning on their own – if a good extension system is not in place.

A wide variety of governmental agencies, such as those that provide finance and help mitigate risk among firms, those that specify and enforce biotechnology-related laws and rules, those that enable parties to contract and conclude agreements, all play a key role. Finally, agencies that represent collective demand such as farmers associations and collectives are a critical set of actors. These main actors and the linkages between them are represented in Figure 2.2 below.

Both the interdisciplinarity of the scientific and knowledge base as well as the complex processes involved in bringing products from the firm to the farm make a range of knowledge interactions critical to competence-building at the sectoral level:

- Knowledge interaction between university departments, centres of excellence, and public research institutes (PRIs).
- Knowledge interaction between traditional knowledge holders (farmers communities) and other more research-based and product development actors.

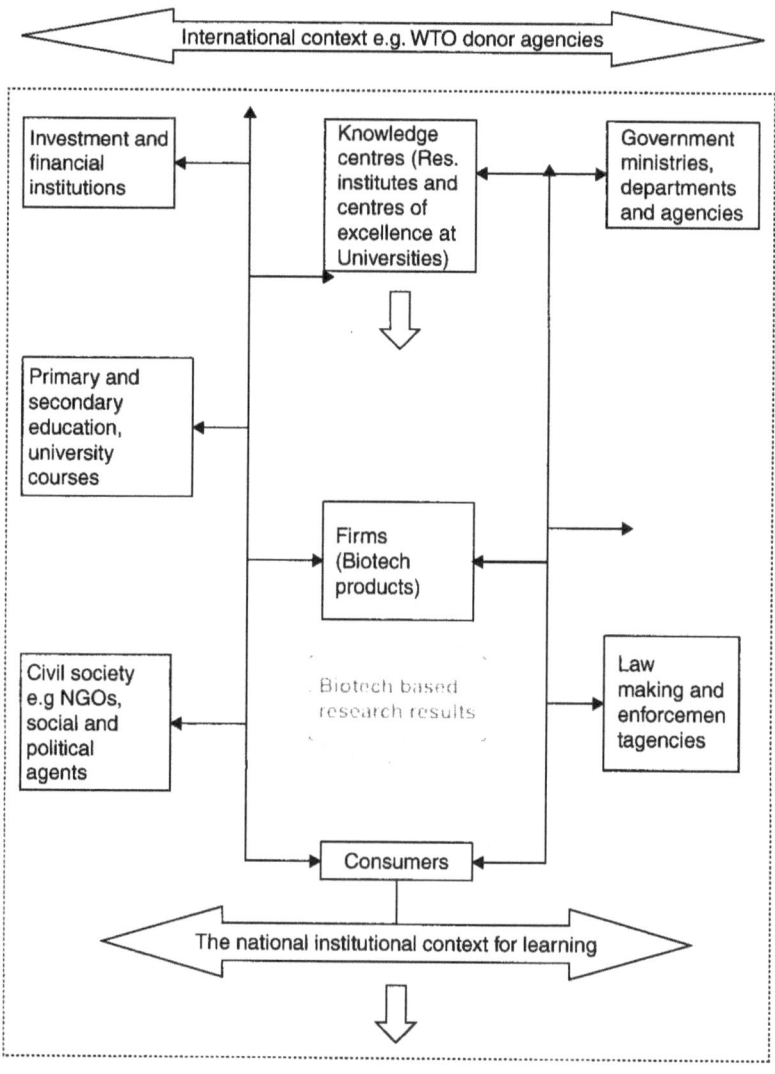

Figure 2.2 Biotechnology sectoral system of innovation in developing countries
Source: Authors.

- Knowledge interaction between local and foreign firms and universities.
- Knowledge interactions between local and foreign firms and PRIs.
- Knowledge interactions between local and foreign firms.

- Interactions between farmers, consumers, seed banks and other intermediary organizations that help gauge local demand and issues imminent to the agricultural system.
- Interactions between various governmental agencies responsible to promote these competencies locally.
- Types of interactions, and within what disciplines of science.
- The determinants and variety of knowledge interactions.

Such knowledge interactions are difficult to measure, and can be done through a composite of factors that include:

(a) the percentage of monetary allocations devoted to research;
(b) the percentage of research contracted to and from outside organizations;
(c) level of joint research with other organizations (basic, applied, or product development initiatives);
(d) the number of scientific publications jointly written with other institutions;
(e) level of co-authorship based on joint research;
(f) exchange of key technical and scientific personnel (numbers and levels of qualification);
(g) involvement in joint R&D programmes organized by the government at the sectoral level, and
(h) consultancy research carried out for other organizations, both local and foreign.

Our surveys in all countries analysed in the subsequent chapters were aimed particularly to capture these forms of interactions.

We map the agricultural biotechnology sector by identifying *sector-specific organizations* split across five essential components:

1. *The knowledge component:* This primary involves formal research organizations producing mainly codified knowledge, mainly in the public sector, but recognizes that the private-sector research and training organizations.
2. *The enterprise component:* This largely involves firms and mainly involves using codified and tacit knowledge and producing tacit knowledge.
3. *The demand component:* This primarily involves consumers and domestic and international markets for products. It also includes policy actors while these are not consumers in the conventional

sense. However, they have a demand for knowledge and information produced by the innovation systems (to inform policy) and need to be thought of as an integrated part of the systems in just the same way as consumers of more conventional products.

4. *The support systems component:* Organizations in this domain may not necessarily be involved in creating or using knowledge, but they play a critical role in ensuring knowledge flows form one part of the systems to another. This might involve articulating demand for knowledge or products from disadvantaged or fragmented constituencies such as SMEs. This would include standard and quality control organizations, trade and manufacturing, and industry associations. Alternatively it might be organizations that make a business out of brokering access to knowledge. These might be consulting companies, or third-party agencies such as those trying to provide developing countries with access to new technologies.

5. *Political, legal and regulatory component:* Organizations and institutions that make the rules and enforce them.

This typology is far from perfect. The categories are not mutually exclusive. Actors can play multiple roles and these roles can evolve over time (see below). Nevertheless it provides simple guidance on the sorts of organization that are likely to be important in a sectoral innovation system.

2.3.2 Institutions supporting collaborative learning

Collaborative learning is pivotal to innovation capacity due to the dynamics of biotechnology-based research and production, which thrive on not just the presence of a strong science system but the exchanges between organizations. Three features that characterize the agricultural biotechnology system makes collaboration imperative, namely, the increasing costs of doing innovation, the increasing multidisciplinary of scientific research in disciplines such as chemistry, agricultural sciences, livestock research, generics and molecular biology, microbiology and biotechnology, and the narrowing of the gap between basic research and industrial application (Kaiser and Prange, 2004).

Since innovation processes are heterogeneous, factors and policies that may trigger off optimal interactions between various systemic counterparts vary from one country to another. The schematics in Figure 2.3 capture some of the triggers to collaboration incentives which feed into the system from multiple sources, namely, international and national policies, finance institutions, physical infrastructure and extension

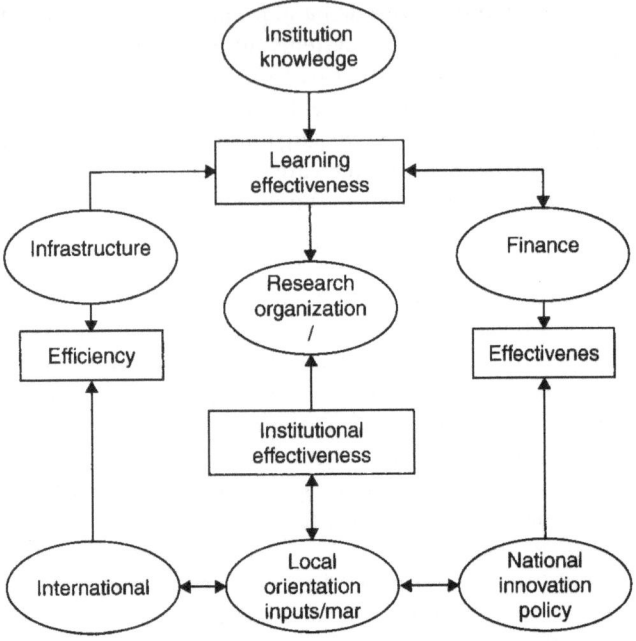

Figure 2.3 Institutions supporting collaborative learning in agricultural biotechnology

services, and local market-orientation of research and products. A range of externally imposed factors, such as the multilateral trade regime and intellectual property rights, play a large role in determining the ways and means in which innovative capacity is built, sustained, and deepened over time in developing countries. Latecomer countries that succeed in attaining competitiveness in agricultural biotechnology demonstrate extensive policy competence in balancing local and international convergent factors that shape systemic learning.

2.3.3   Innovation capacity of the system

We identified the four core components of innovation capacity in Chapter 1 to be:

1. The ability of firms and organization to undertake product-related, imitative activities.
2. The ability of organizations to build competence (e.g. universities and PRIs).

3. The ability of institutions to eliminate distortions in information exchange which hits at the roots of the missing or ill-formed collaborative linkages in latecomers.
4. The ability of the state to formulate policy.

Latecomers and very latecomers exhibit substantial differences in these institutions from countries at the frontier and these differences provide four key factors – namely, interactive learning, knowledge bases for biotechnology, incentives for translating inventive activities, and state policy capacity for defining and comparing innovation capacity in agriculture biotechnology. In the remaining four sections, we evaluate the differences in each one of these capacity elements in fast followers, latecomers, and very latecomers and how these impact on their sectoral systems of innovation for agricultural biotechnology.

2.4   Knowledge bases

As the nature and extent of knowledge is pervasively linked to actors and institutional arrangements, we conceptualize knowledge crea-tion, validation, and dissemination as a social process where groups of scientists and engineers could be seen as a community (David and Foray, 2001). Agricultural biotechnology has a very strong knowledge dimension and it builds on two potential networks or communities engaged in this process. The first is the research-intensive public sci-ence, organized largely around those creating new knowledge through intensive R&D activities and creative design that initiate entire proc-esses of innovation. Although it is difficult to draw the lines conclu-sively, basic research (and some applied) tends to be the domain of universities and highly advanced public laboratories while firms tend to focus more on applied and developmental research. This was largely led by the public-sector organizations even in the frontier countries, up until recent times, because private firms tend to have little incentive to engage in socially relevant basic research due to problems of appro-priability and uncertainty, as captured by the entire discussion on the reasons and use of intellectual property as an incentive for knowledge creation. In recent times, trends that promote the commercializa-tion of university/public-sector research have blurred the boundaries between basic and applied research, as well as tending to promote the privatization of some basic research. The specialization of actors as well as complementarities involved is further demonstrated through some examples. The United States spent an estimated US$54 billion on basic research, US$66.4 billion on applied research, and US$187.3 billion on development in 2006. In proportional terms, these are 18.7%,21.3% and 60.0% respectively of total R&D spending. Out of these, private firms in the US spend three times more on *applied* rather than on *basic* research. Industrial research is in fact dominated by developmental research, which accounted for 90.2% of development work carried out in the country in 2004; universities and related actors spend less than 2% on development research (NSF, 2006).

The second is a set of actors driven largely by commercial motive of translating inventive or design work into products, processes, and services; they are made up in the main by engineers, scientists, software experts, designers and technicians, and practitioners such as farmers. Both these communities may overlap at times, and they are found in networks created to advance technological innovation. In the context of innovation studies, what underlines these two communities is the commonality of knowledge creation.[8]

The knowledge base of both these networks comprising farmers, producers, and organizations grows on the basis of routine learning-by-doing where the individual capabilities of actors are largely tacit. However, a firm/organization cannot rely indefinitely on its own internal capabilities developed at a particular time without the periodic injection of new ideas, whether through internal R&D or acquired from external sources. This is the imperative of innovation. Therefore an important point for theory and policy for supporting industrial produc-ers in late development is to understand how the tacit knowledge base in industry and public-sector organizations grows and how firms and organizations could best transform codified information into relevant knowledge.[9] Tacit knowledge grows on the strength of these formal channels but more directly on other informal channels such as doing reverse engineering, interactive learning through collaborative R&D with other actors, skills exchange through personnel movement, con-sulting technical and trade journals, participation at expert meetings, and apprenticeship training.

Universities and PRIs are *loci* of knowledge creation within systems of innovation in biotechnology. The role of universities in this ever-changing political, social, and scientific landscape that facilitates the growth of new technologies has been evolving, with a closer, more structural relationship between universities, public laboratories, and industry becoming a central part of the capability-building exercise (Meijer and Witling, 1997). There is a self-reinforcing relationship between how universities evolve to become centres of knowledge generation and collaborative networks in biotechnology – on the one hand, research excellence makes the institutions promising partners for collaborative ventures with industry, whereas, on the other hand, collaborations reinforce the strengths of universities in achieving research excellence. But the policies and practices of a country with respect to its universities are shaped mostly by the country's singular historic development and the social factors that are institutionalized over time (Bartholomew, 1997, p. 244). These play a pivotal role in determining institutional behaviour and cooperation patterns between universities and other counterparts in the biotechnological systems of innovation. Factors that motivate inter-organizational collaborations are not only localized, they can differ radically from one social and institutional context to another. If this is true, then the capacity of universities to evolve as centres of knowledge excellence may be specially challenged in social and institutional contexts that are prone to limitations and sporadic interventions. In such environments, determinants of collaborative strengths may not only be different from what is witnessed in the frontier countries, where purposive interventions are possible, but systemic and institutional weaknesses that stem from historical reasons may also create perverse incentives for knowledge generation, use, and transfer in biotechnology.

Three kinds of firms would appear to be the active components of a biotechnological sectoral system of innovation: established agricultural companies (for seeds, pesticides, and other agricultural inputs), independent biotechnology firms, and firms performing a wide variety of activities that can be clubbed together as 'knowledge-intensive business services'. While we do not need to dig deep into the firm-level characteristics of large agricultural companies, the firm-level competitiveness of independent biotech firms is incumbent upon their ability to access and make efficient use of markets for technology, on the one hand, and collaboration networks, on the other (Allansdottir et al., 2002). It is not easy to sustain the growth of independent biotechnology firms in the agricultural sector: data from the EU shows that the amount of new firms entering the agricultural biotechnology declined gradually from 15% of all biotech start-ups in 1995 to just 5% of all biotech start-ups in 2000. The firms that provide knowledge intensive services are critical for the sector because of their ability to perform:

(a) knowledge-intensive work;
(b) consulting services – in the legal, marketing or other domains, and
(c) client-related services.

2.5   Interactive learning and innovation capacity at the sectoral level in latecomers

Interactive learning is perhaps the first key characteristic of innovation capacity of the agricultural biotechnology sectoral system: the nature and speed of the interactive learning processes are germane to the way knowledge is generated and absorbed. Collaboration therefore as the precursor to innovation is underpinned by institutions that are designed to be carriers of knowledge, representing the cumulative learning of groups and societies (North, 1996). The 'memory' represented by institutions is particularly important for tacit, non-codified knowledge. The speed of economic change is a function of the *rate of learning*, but the direction of that change is a function of the expected payoffs to the acquisition of different kinds of knowledge (North, 1996, p. 346, emphasis added).

Three aspects of agricultural biotechnology make collaboration extra significant. First, that it is an assortment of interdisciplinary techniques that are cross-sectional in scope and generated significant discontinuities from extant modes of production and, for this reason, established norms of scientific practices are disrupted and machines are rendered obsolete. Second, the new scientific and technical regimes it brings into use call for new institutions and organizational structures and norms. Third, established organizations and firms may be forced to adjust to the new conditions or lose out to newcomers as it has been with the emergence of new technology-based firms (TBFs) in the global biotechnology industry.

2.5.1   Universities as centres of excellence

The association of technical skills and institutional capability with performance is well documented in innovation literature. The proportion of in-house graduates and technical skills is a proxy and capability and relies on the availability of educational institutions at the secondary and tertiary levels that provide good quality education of relevance to the sector. The intensified knowledge-oriented environment tends to require not only a higher level, but also a wider range of skills (Lall, 2001). Universities not only generate new knowledge, but also produce skilled manpower through teaching, public research, and development (both basic and 'other types of pre-competitive') that are primarily aimed at correcting market imperfections (Drejer and Jorgensen, 2005). As Schartinger et al. (2002) elaborate, universities are meant to play three distinct roles in the knowledge environment:

- they conduct fundamental and applied research that shift the knowledge frontier
- of industry over time;
- they generate innovations that are of immediate relevance to industry, and
- they produce human capital through the training of scientists, engineers, and researchers.

While firms themselves do have internal R&D facilities, external knowledge is an invaluable source of firm capability.[10] The role of public research has in this manner been pivotal to the advances made in industry in general, and biotechnology in particular as the evidence presented from 15 EU countries earlier on in this chapter shows. Biotechnology's locus of creation lies entirely within universities thereby underscoring the importance of university–industry collaboration for biotechnology-based innovation for any sector. In the US, where it first evolved, researchers in academia made the most significant biotechnological discoveries. Elsewhere too, greater specialization and concentration in research activities and increasing competition among biotechnology companies forged the need for greater links between academia and industry (Meijer and Wilting, 1997).

The final important function of universities is the production of trained manpower and the mediating role of personnel exchange between academic institutions and industry. This important human capital function relates to the general mobility of scientists from universities to industry, and through a variety of collaborative arrangements between universities and PRIs and industry, which is discussed at length in the next section. The skills market in latecomer countries faced persistent market failures, which is not only limited to skills formation, but also access to relevant information, finance, and technology markets. In biotechnology, not only do latecomer countries lag far behind in introducing the state-of-the-art courses (not to mention the dismal status of the laboratories in which students perform laboratory experiments), but also, most courses relate mainly to the S&T aspects of biotechnology and do not span to other areas, such as marketing and legal aspects. These result in huge manpower shortages for commercialization efforts.

2.5.2 Firm–university inter-linkages in agriculture biotechnology

There are a variety of reasons why firms look to create linkages with other actors and why organizations themselves increasingly seek to commercialize their inventive activities. First, firms and organizations increasingly seek external knowledge because of the growing complexity of production (Howells, 2000) and, second, they do so because of the interactive nature of learning (Lundvall, 1988; Freeman, 1987). In biotechnology where research and application take place almost simultaneously, the role of universities assumes added importance within the system of innovation (Fontes, 2004; Orsenigo, 1989).

There are several advantages of university–industry PRI interfaces in biotechnology. These are:

1. Lower R&D costs for the companies: Recent research conducted on university–biotech firms alliances shows that such collaboration is critical to bring down R&D costs of companies while achieving a higher innovative output.[11]
2. For the universities, collaboration enhances sources of public and private funding as well as helps partake in technology transfer and human training activities. It also helps enhance focus on secondary research of immediate industrial relevance.
3. Universities interact with PRIs perform a supportive role in two stages of biotechnology-based innovation – research and patenting. In research, universities collaborate with PRIs and biotechnology firms in providing the requisite knowledge base to the industry or higher research institutions. At the same time, they also utilize the supportive services provided by PRIs, to structure research more towards industrial applications.

The literature on public research and university–industry interaction is considerable and scholars address different aspects of the transfer process.[12] We attempt to present the main results that accrue from this literature as a set of stylized facts here. First, the nature of knowledge generation and transfer is complex, highly systemic, and context-specific, particularly as a result of the significant but hardly acknowledged tacit content of scientific skills required. These kinds of tacit knowledge are not so easy to set out. Tacit knowledge refers to this bundle of information that is more embedded in the skills of the performer, aspects that are more easily expressed because it is built from considerable practice and accumulated experience. There are many dimensions to tacit knowledge,[13] but much of the tacit knowledge in firms is transformed into organizational routines (Nelson and Winter, 1982).[14] Routines are regularities and predictable patterns of behaviour. In small firms, the owner/manager tends to define and exemplify the nature of routines. In apprenticeship institutions, the master personifies the routines and determines the culture and rates of transferring this largely 'hard to pin down skills' to learners. Similarly, in other organizations, organizational routines are the determining parameters in the acquisition of tacit knowledge. Therefore, countries need to develop domestic scientific capabilities suited to their own context in order that both codified and tacit skills are locally available. Second, there is a wide gap between the motivation, scope, and purpose between academic research and industrial research and production. This complicates the transfer process, and restricts the scope for policy incentives (Dasgupta and David, 1994). Third, firms seek external collaboration for purposes of learning because autonomous efforts are costly, and innovation outcomes are uncertain. Moreover, collaboration releases firms from additional financial commitments, thereby allowing firms to share risks and spread sunk costs (Bougrain and Haudeville, 2002). Whereas, learning results in new ideas from combining experiences (Hakansson, 1987), inter-firm and inter-organizational cooperation results in the exchange and dissemination of knowledge (Teece et al., 1990) and hence are very important in promoting innovation.

2.5.3 Knowledge interactions and collaboration

In order to translate local demand and inventions to commercially viable innovations, the onus is on institutions that can foster collective learning. There is a range of institutional interventions that provide ancillary mechanisms for learning to proceed, both in the form of generic national innovation policy instruments and in the form of dedicated biotechnology policy instruments. Dedicated biotechnology policies, however, rely on the pre-existence of a strong science base, which is in itself a questionable attribute of latecomer development. These need to be preceeded by generic policies that clearly promote the science base, including (Enzing et al., 2008):

- Policies that fund basic research (to promote high-level biotechnology research in universities and PRIs).
- Policies that fund applied and industrial research (to promote industry-oriented work in PRIs).
- Policies that make available human resources (for the development of the sector).
- Policies that create dialogue with the public (to steer public perceptions of biotechnology into constructive engagement).

Empirical research on several countries in the EU (Enzing 2004 and 2008) shows that countries with successful biotechnology sectors have a good mix of both generic and dedicated biotechnology instruments.

2.5.4   Latecomer specifics in enabling interactive learning

However, in a latecomer context, collaborative learning does not just happen because there are considerable obstacles to interaction among actors within and outside national borders. First, the nature of knowledge generation and transfer between, for example, universities and PRIs, and industry is complex, highly systemic, and context-specific, particularly as a result of the significant but hardly acknowledged tacit content of scientific skills required which will therefore require more than codified format. Second, there is a wide gap between the motivation, scope, and purpose between academic research and industrial research and production. This complicates the transfer process, and restricts the scope for policy incentives (Dasgupta and David, 1994). Third, firms and organizations tend to explore external collaboration for purposes of learning because autonomous efforts are costly and innovation outcomes are uncertain. Firms therefore would focus on core activities and prefer to specialize; moreover, collaboration releases firms from additional financial commitments, thereby allowing firms to share risks and spread sunk cost. However, such learning that could result in new ideas from combining experiences ignores elements of heterogenous agendas and conflict of interests between the public- and private-sector enterprises that are prominent among latecomers.

Fourth, most small firms do not have formal R&D structures and considerable but largely unrecorded changes do take place at the shop-floor level and identifying and recording R&D investment is therefore quite problematic. The current internal S&T capability of firms is associated with the ability of the firm to exploit external knowledge while extant capabilities result from previous investments made by firms. In short, a firm's ability to take advantage of market opportunities through innovation is path-dependent. As Bougrain and Haudeville (2002, p. 746) note, 'internal expertise facilitates the identification of external information, their absorption and improvement of SMEs' performances'. This indicates that the nature of knowledge activities are predominantly in the incremental design realm and not in the R&D-based domains (Oyelaran-Oyeyinka and Gehl Sampath, 2009) as would be required for biotechnology. Fifth, despite the huge investments on PRIs, they are often ranked low as sources of information. For instance, in most of our empirical surveys across over 15 latecomer countries in different sectors, less than half of the firms found the importance of government laboratories to be either moderate or very significant.[15] No firm indicated that the information from universities or government laboratories was crucial for the innovation process.

There are three specific sets of issues that limit interactive capacity among latecomers – which we set out here and also capture through our empirical work to substantiate our findings. The first of these is *inter-organizational limitations* could stem from the weak institutional basis that organizations tend to have in latecomers – mostly, organizations have no competence building power within themselves because they are established without mandates, or without adequate funding, or have overlapping structures or have no substantial human capacity to draw upon in order to function in an effective and outcome-oriented way. Organizations tend not to have the power/mandate/manpower capacity to create new generations of academics, scholars, and scientists that can partake in the process of knowledge-led economic development, which manifests in innumerable ways, including weak industry–public-sector linkages. Second is what we call local network factors. The local network factors tend to reflect in the resource-constrained environment for innovation. There is an extreme symbiotic relationship between the national innovation system and the sectoral system – a large number of the institutions and incentive instruments that promote interactive learning span across sectors and are generic, and when these are not well-structured, enacting sector-specific institutions may not go a long way. Whereas fast followers have more skilled professionals, better public research infrastructure, good enterprise sector, relatively good policy competence, functioning risk instruments, and collaboration incentives, latecomers and very latecomers suffer from the lack of these very basic framework conditions. Finally, *global–local knowledge factors* refer to the external impetus to local biotechnological activities that stymie the perception and real-time activities related to knowledge creation by way of intellectual property rights and breeders' rights, sanitary and phytosanitary measures, biosafety requirements, and public perceptions of genetically modified (GM) technologies. While there are a few positive global–local linkages, such as global value chains in horticulture and agriculture production, these seem to be overshadowed at times by the negative implications of breeders' rights, as our case study of the cut-flower sector conducted in the chapter on Kenya in this book shows.

2.6   Rough road to the market

The most significant innovation capacity constraint among latecomers
lies in translating research into innovative products. In other words, it is
not so much that there is no base to draw upon for innovation, but most
assumptions of how individuals and organizations interact do not hold
good and the process of translating R&D efforts or incremental inventions into a market product is characterized by significant financial, skills,
technical, and institutional barriers. Such barriers are often all the more
observable in research into and translation of new technologies such as
biotechnology and information communication technology (ICT) into
marketable commodities since it requires even greater resources. The
path from the laboratory to the market could be long and expensive,
and the outcome highly uncertain (see Figure 2.4). Institutional and
structural factors pose significant and unexpected obstacles to collaboration, particularly in science-based sectors that are characterized by idiosyncratic technical and scientific properties. However, failure to achieve
optimal coordination through appropriate institutional incentives will
only result in sub-optimal outcomes in least developed countries, and
merely investing in R&D is not enough. However, failure to achieve optimal coordination through appropriate institutional incentives will only
result in sub-optimal outcomes in least developed countries, and merely
investing in R&D is not enough, as figure 2.4 below shows.

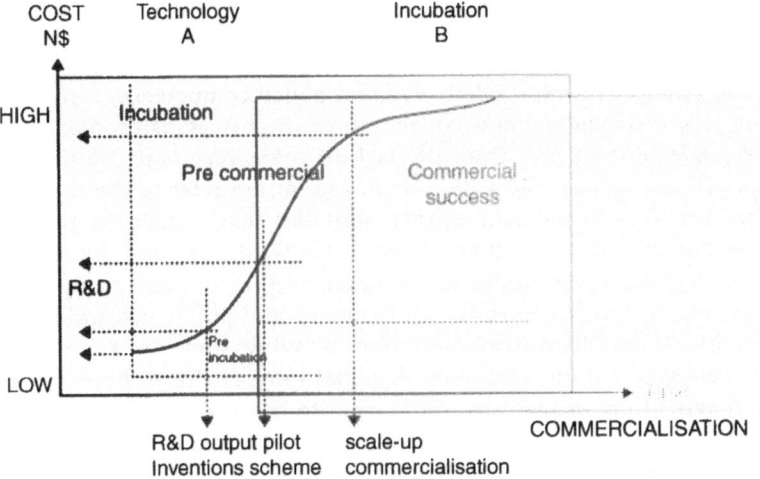

Figure 2.4 Illustrative graph of the financial demand of scaling R&D outputs
and/or invention

Regardless of whether the initiating step is the incremental design or laboratory-based R&D stage, these factors are critical determinants of commercialization. There are three clear inter-relationships that prove to be the reason:[16]

1. Infrastructure (both physical and knowledge) has strong correlation with innovation performance.
2. Interlinkages between key agents in the domestic knowledge system, as promoted by dedicated biotechnology instruments for collaboration has strong correlation with innovation performance.
3. Government policy on innovation (both generic innovation framework and dedicated biotechnology policies for finance) has strong correlation with innovation performance.

We have already examined the physical and knowledge infrastructure in the foregoing sections, and therefore will focus only on the second and third determinants of product development here.

Biotechnology commercialization experience among frontier countries employs three kinds of dedicated policy instrument (Enzing et al., 2004):

(a) *Those that support the commercialization of university research and 'spin-off' scientists.* Given the science-intensive nature of the sector, these are often extremely essential in both promoting mobility of scientific labour between university and industry and also ensuring that university research is more outcome-oriented. Policies that promote this include grant of intellectual property rights (IPRs) to universities, geographical clustering policies that locate biotech clusters around science-based universities, technology link foundations, and biotechnology exploitation platforms.

(b) *Those that reduce risk of innovative activities through finance.* Finance, given the uncertainty of the outcome of translating R&D efforts and inventive activities into successful commercial ventures, is central to product development initiatives. Normally, across sectors, funding demand for the scale-up of R&D and inventive efforts increases exponentially from prototype to commercialization and the success rate of scaling-up outputs and/or inventions is generally low (less than 20% in US and Canada, more than 50% in China where government support is most significant) even in the best of circumstances (defined in terms of adequate physical and knowledge infrastructures, technology, and funding). Support models to help agents overcome financial constraints and engage in biotechnology innovation include government-support soft-loans,

R&D subsidies, public risk capital funds, public support for private enterprise through grants, subsidies, private equity, venture capital, and buy-out investments. Policies that could promote these (from a review of literature) include: seed-financing programmes, Business Angels networks, enterprise subsidy programmes (for setting up new biotechnology start-ups), common placement funds for innovation, and research tax credit programmes.

(c) *Those that provide other forms of business support to fledgling ventures.* Emerging biotechnology entrepreneurs require a range of specialized skills that are very important during their early stages of growth (Enzing et al., 2004). They battle with issues related to intellectual property protection (what is protected, what not, how to obtain licences), marketing and demand assessment, advertising and reaching consumers, among many others. These call for policies that provide technology incubation facilities, S&T parks, competence centres that provide expertise such as legal affairs and marketing.

Government policy, in terms of the institutional framework, impacts upon innovative performance in a monolithic way. Institutions are defined as the 'rules of the game', they imply routine behaviour and actions and consist of the 'cognitive, normative, and regulative structures and activities that provide stability, coherence and meaning to social behavior' (North, 1996). The concept of 'institution' is often used interchangeably with that of 'organization'. While the normative and cognitive aspects of institutions are stressed, greater emphasis is laid on the structural dimension of organization. We follow the definition of an institution provided by (North, 1996): 'Institutions are the rules of the game of a society or more formally the humanly-devised constraints that structure human interaction. They are composed of formal rules (statute law, common law, and regulations), informal constraints ... and the enforcement characteristics of both'. Institutions being far more pervasive and often more influential in their impact than the economic system, tend to exert profound effects on the internal processes of producing entities. Herein lays the importance of institutions in determining the rate and direction of technical change.

The role of government is therefore critical in as much they are the custodian of policies and institutions. More serious for the latecomer, however, is the evident reality that 'backwardness has been relatively greater' requiring even greater dose not just of state action but of competence. These manifest in a number of ways: the absence of strong and competent state institutions, weak entrepreneurial business firms,[17] relatively low level of skilled engineers and managers, and well-educated and abundant low-cost managers (Amsden, 1989; Amsden and Chu, 2003). Although Gerschenkron (1962) had observed that in catching up latecomers would potentially have access to a basket of proven technologies, much of what development requires are in the realm of tacit knowledge with significant preconditions for interpersonal learning. In other words access to this source is not automatic and institutions that mediate knowledge acquisition whether markets or networks are in large part absent in developing countries. Lastly, as Amsden and Chu (2003, p. 13) observe, market forces are unkind to the weakly organized economies ('the more backward the country, the harsher the justice meted out by market forces') with its inherent and often contradictory requirements. The role of governments is therefore critical in as much as it is the custodian of policies and institutions as well as the means of enforcement.

An economy is made up of a wide variety of economic and non-economic actors that are coordinated within a system involving both markets and non-markets transactions. Coordination is effected by means of institutions that are both formal and informal (rules, norms, social conventions, and statutes). Where these institutions are missing or inadequate, the role of industrial and innovation policies is to provide institutional compensation, a point which is discussed at length in the next section.

Institutions for innovation are extremely important to long-term economic growth precisely because technology mediates the introduction of new products and processes in the economy. In an industrially dynamic context, changes to the machinery and equipment, the introduction of new forms of industrial organization will be accompanied by 'new institutions, the institutionalisation of ... new social technologies may require new law, new organizational forms, new sets of expectations' (Nelson and Sampat, 2001, p. 49). The corollary is that, in a situation of economic backwardness, changes to institutions are rare just as technological innovation might be equally rare or non-existent.[18] Institutional changes become even more crucial in a time of human engineered structural change.

Institutions in latecomer contexts exhibit ineffectiveness in establishing efficient inter-linkages and incentives for agents to engage in learning and knowledge creation activities. Their inefficiencies gives rise to the poor coordination of knowledge and economic production functions, leading to imbalances in the demand and supply for skills of the right kind, quantity and quality mix at sectoral levels and over time. A major reason for this lies with government involvement that tends to create its own idiosyncratic lock-in conditions for two main reasons. First, instead of governments playing a supportive role to rectify market imperfections, governments in latecomer contexts have over time lent strength to the creation of institutions that override market forces, thereby creating alternative institutions to which actors have to respond, albeit to promote self-interest-based, inefficient outcomes. This capture of the entire institution-building process is a commonplace occurrence. As a result, while there is a general agreement that developing countries need to create organizations and institutions where they do not exist and reform those that are functioning poorly, institutions for policymaking themselves lack both broad and specific competencies in their coordination functions. This is a serious drawback for developing countries and leads to a situation in which policy coordination is largely politically driven in the absence of strong market coordination.

Other institutional gaps due to lack of funding of relevant organizations only attenuate the consequences. Maintaining these organizations to achieve effective service delivery depends to a considerable extent on available public resources. But poor financial commitment for meeting organizational commitments, results in disillusion of scientists and researchers over time, lack of private-sector trust in collaborations with public-sector institutions, and often even prevents the rise of a private sector itself since organizations to promote the growth of private-sector firms do not function effectively.

2.7   The role of the state in promoting agricultural biotechnology

The state plays different and sometime overlapping roles, which broadly include that of the custodian (regulatory) and producer or both. The custodian function is based on the presumption that private capital is limited and inadequate in its engagement with the evolutionary assignment of industrialization.[19] This role of the state will depend as much on the technological knowledge domain as it is on sectoral differences. According to Evans (1995, p. 11), 'What roles states try to play depends on each technological and organizational characteristics of the sector. How well the roles are played and with what consequences depend on each state's institutional characteristics'. In other words, there are wide variations in the attributes of the states and therefore 'states are not generic' (Evans, 1995, p. 11). In very significant ways, states are constituted of diverse structures, which in turn create different capacities for the task of transformation.

This notion that structures determine state capacity resonates with the idea that countries are differentiated by their innovation capacity.

The kernel of the development state thesis is that the state should define and set long-term goals for transformation and structural change to induce growth and development; and, as well, put in place mechanisms to manage political relationships and complex actor interactions. We have examined the role of the state more generally in advancing knowledge growth and innovation (Oyeyinka and Gehl Sampath, 2009), and here we specifically identify five roles for the state in the pursuit of technology and innovation for agriculture, namely:

1. States need to formulate a long-term vision and goal and pursue it.
   There has been much focus on the imperative of state-vision in technology-led development, especially by scholars who have analysed the East Asian successes. Both Freeman (1987) and Johnson (1982) allude to the technology imperative that dominated Japanese long-term development goal.[20] The catch-up path of both Korea and Taiwan were equally driven by a technology vision (Amsden, 1989; Amsden and Cho, 2003; Mathews, 2000). Japan set out to develop decidedly on a technology vision and as Freeman (1987, p. 35) notes, 'MITI and other Japanese ministries saw it as one of their main responsibilities to encourage the introduction of new technologies through new investment'. Clearly, state policy has a very important role to play in optimal resource allocation for the society as a whole.

   In the case of agriculture, a vision to achieve self-sufficiency in food grains and enhancing crop productivity have been the drivers of the Green Revolution among those countries who successfully achieved it, and are following it up with biotechnology-based research programmes.

2. States have to provide coordination in order to bring harmony and efficiency to the action of multiple actors acting in a system.[21]
   Biotechnology regulation is not just about biosafety rules and the capacity to enforce them. As we have noted earlier, policy frameworks call for enhanced policy capacity in order to design and implement innovation policies that provide a critical science base, collaboration incentives, risk minimization associated with commercialization of products, extension policies that eliminate information asymmetries among farmers, promote uptake of new locally useful varieties, enhance farm-level production capacity, and access to markets. Very

importantly, the policy framework has to balance these needs and provide incentives for all actors to act conjointly.

There has been extensive focus on the need to regulate biotechnology and GM technologies from a biosafety perspective in latecomer countries, but what should not be ignored is that a lot of this debate stems from Europe's own past experiences on this issue. While so much is being made out of GM crops and their environmental impact, two critical facts remain obscure from policy debates in latecomer countries. First, the entire discussion on GM crops in Europe has predominantly been shaped by the need to control companies to avoid their previous experiences and pitfalls of pesticide regulation in the 1970s, especially with relation to organochlorine. Chataway et al. (2006, p. 172) note: 'As the companies developing GM crops were those that had previously produced organochlorine and other pesticides, it is not surprising that they became the first industry sector to be systematically subjected to a precautionary approach to its regulation'. Second, the political positions of countries are being enforced by way of sanitary and phytosanitary measures (for the protection of human health and the environment), which puts immense pressure on latecomer countries not to adopt GM products for fear of loss of export markets. In particular, the very strict guidelines followed by the EU countries has caused a flurry among latecomer countries wanting to conduct field trails on important varieties of GM crops, that could be locally relevant, out of fear that contamination of export-oriented crops could lead to their eventual rejection and related losses. Scholars have repeatedly warned that the policy guidelines provided by the Protocol, on which countries are to structure their biosafety laws could affect biotechnology development *per se*.

For biotechnology development, the range of policies that latecomer countries need to enact and balance is listed in Table 2.2. The requirements of the TRIPS Agreement especially those related to protection of life forms, as part of its article 27(3)(b) have been particularly challenging for agricultural biotechnology in latecomer countries. Article 27(3)(b) provides a choice between patents and a *sui generis system* (a system of its own) for plant variety protection. Several frontier countries provide both patents and plant breeders rights to plant breeders and the Union of Plant Varieties (UPOV) has been in force since 1961 and has had three main conventions that provide regimes for the protection of plant varieties in 1972, 1978, and 1991.

Table 2.2 Policy capacity for agricultural biotechnology

Policy	Function	Intended result
National science, technology and innovation policy	Provides the basic framework for innovation in the country Provides coordination mechanisms between various systemic component	Promote technology-led development in all sectors of the economy
Dedicated biotechnology policies	These build further on the national innovation framework and provide additional measures specifically required to promote the growth of the sector	Biotechnology research and product development
Industrial Property Act	Grants IPRs, specifically patents and trade marks	Promotes investment and protection of innovation
Biosafety regulations	Ensures human, animal and environmental safety	For the safe transfer, handling and use of GM organisms
Food, drugs and chemical substances regulation	Protects against the adulteration of food and drugs	Sets standards for food, chemicals and drugs
Standards regulation	Sets standards for quality, purity and labelling	Standards setting, verification and implementation of codes of practice (CoP)
Regulations on biotechnology, biodiversity and genetic resources	Ensures balance between the provisions of the TRIPS Agreement and Convention on Biological Diversity for biotechnology and sustainable development	Prior informed consent, protection of traditional knowledge and biodiversity protection
Environmental Management and Coordination	Ensures environmental safety	Provides for environmental impact assessment (EIA)

Source: Authors.

Several latecomer countries have adopted the UPOV 1978 and 1991 conventions as their *sui generis* regimes, regardless of their impact on their local contexts. Literature on the subject has been particularly vocal about the UPOV 1991 convention not containing provisions to cater to the needs of farmers in latecomer countries, and expressly prevents them from saving seeds to sow back in subsequent seasons. At the same time, several other latecomers have devised their own *sui generis* regimes that cater to their own particular contexts.

Providing intellectual property regimes at the national level that take into account specific social, cultural, and public-interest needs of countries and balancing these requirements with local learning and innovation aspirations calls for highly evolved policy capacities in latecomers.

3. States have to put in place institutions where they are missing and strengthen those that are weak (fully discussed in following sections). These institutions include those that foster interactive learning through systemic coordination. Incentive systems tend to develop from more fundamental institutional roots such as labour laws and even national constitutions. Terms of employment and work environments, both tangible (research and teaching facilities) and intangible (possibilities for institutional collaboration, quality of networks and colleagues) play a pivotal role in retaining skilled professionals.[22] States have been involved in promoting academic–industry exchanges by encouraging channels of learning, such as: joint publications, mobility of scientists and engineers, cooperative R&D, facility sharing, research training (e.g. capacity development at PhD level, international and local exchange of staff), IPRs (licences, patents, copyrights), and academic entrepreneurship (spin-off firms) (Brennenraedts et al., 2006).

4. States act as guarantors of risks and provide innovation 'insurance'. It is not uncommon that entrepreneurs are slow to uptake innovation prospects coming out of the activities in a sector/economy due to risk and uncertainty, especially in sectors and technologies that are new to the local contexts. In such cases, successful state action involves the creation of several mechanisms, including newer systems of property rights that insure rents for having taken up the risks of engaging in innovation, as the previous sections of this chapter show. The case of data exclusivity in pharmaceutical and biotechnological innovations is one such example. The EU began granting data exclusivity as a mechanism of protection of rents for biotechnological

innovations at a point of time in Europe when biotechnological patents were not allowed, simply in order to be able to allow entrepreneurs to engage in such innovation activities. Similarly, the grant of plant breeders' rights for agricultural varieties is a mechanism that evolved to guarantee rents for the creation of new plant varieties at a time when there was a need to promote newer varieties to enhance food security in Europe in the 1950s and 1960s. Apart from creating alternate property rights structures, state actions can insure innovation in a variety of other ways, including helping to create or protect consumer markets both local and export-oriented. All these forms of innovation 'insurance' tend to play a very important role especially for firms in emerging sectors in latecomer contexts.

5. States have to manage conflict and resolve problems of asymmetric power relations.[23]

In the face of emerging sectors and industries, powerful actors in an economy with vested interested stand to lose influence, profits and markets: this shuffling is an imperative of change. In other words, some actors might potentially gain while others potentially lose; and more invidious is the uncertainty of the extent of gains and loses. This raises the prospects for conflict. Where then is the platform for bargaining and mediation? In this scenario, the state could act as the overall coordinator with the mechanisms of institutions. In the case of agricultural biotechnology, the state also plays the role of a mediator, constantly creating a platform for dialogue between those who oppose it and those who argue for it based on its merits, in order to pave the way for a balanced consideration of the advantages of agricultural biotechnology in latecomer contexts. This is not to say that the state in question would always have all the requisite capacity and the willingness to undertake such a task. However, the state is in the best position to perform these roles.

2.8 Summing up

We sum up this chapter by calling attention to our original proposition that innovation capacity represents an important shift in the way we understand knowledge generation and use particularly in countries at the catch-up phase. We also call attention to the need to consider a wider set of actors as well as avenues that are non-market avenues; in other words, both hierarchy and networks are important in building a dynamic biotechnology system of innovation. Fragmented markets and poor coordination mechanisms in latecomers are notably weak and will have to be built up. Prominent among these are the structures of R&D, finance support, metrology, standards and quality centres and, at the base of it all, the system of education, which is responsible for new knowledge from basic research and the training of scientists and engineers.

We suggest that building a biotechnology system will entail promoting policy capacity for dynamic innovation policies in a global system that has become more complex, knowledge-based, and innovation-driven. Central to our innovation capacity framework is the fact that policies and institutions matter. An important role of innovation is that it attends to not just the activities of the firm, but the interaction between economic and non-economic actors. We agree with the proposition that *interactions should be facilitated by means of policy-if they are not spontaneously functioning smoothly.* This is a particularly significant function because developing countries lack organizations and institutions for the regulation and coordination of innovation activities.

Finally, this chapter makes the point that there are distinct differences between Science, Technology, and Innovation Policies and that innovation policy should relate the complexities of the innovation process to development objectives of tackling poverty and reducing deprivations such as hunger.

3
Malaysia Biotechnology:
A Fast Follower

3.1 Introduction

Malaysia is an example of a fast follower in our framework in terms of its general performance and innovation indicators. With a third of its economy still dependent on agriculture and natural resources, Malaysia has recognized that biotechnology will enhance the productivity of the sector while creating new opportunities for the emergence of new industrial sectors such as healthcare and industrial biotechnology. The biotechnology industry is relatively new in Malaysia although food and food additives produced by conventional biotechnology such as fermentation processes and tissue culture (TC) have been in existence for decades in the country. Biotechnology and within that, agricultural biotechnology research and development (R&D) was identified an issue of national interest as early as the 5th Malaysian Plan (1986–90), but it was prioritized as a potential lead sector only in the 8th Malaysia Plan of 2000–5.

Malaysia has a wide biotechnology research base spread across its public research institutes (PRIs), universities, and the private sector drawing primarily on its strengths in agricultural research. Companies such as Guthrie, Sime Darby, Golden Hope, IOI, and United Plantation have a long history of the application of TC for agriculture and horticulture. Recombinant technologies and DNA marker technologies have been applied to several plants for identification, inheritance studies, marker-assisted selection in breeding, and genetics mapping. Animal recombinant vaccines have been produced to promote the development of the animal husbandry industry. The application of bioremediation techniques in the treatment of industrial and agricultural waste is already underway, in the country while diagnostic kits for tropical diseases have been developed. In recent years, a wide variety of international partnerships have been established focusing on natural products discovery, agricultural biotechnology, and the expanding niche market for natural and specialty products, in which Malaysia hopes to be a significant biotechnology player.

This chapter presents the results of our survey of 76 biotechnology firms engaged in agricultural R&D in the country, which was substantiated by face-to-face interviews with a series of stakeholders, including governmental agencies, PRIs, and universities in 2007.

3.2   Key actors generating agriculture innovations and knowledge

The current priority areas of relevance to agricultural biotechnology include:

(a) *Plant biotechnology*: which includes genetic engineering for plant improvement, biopharming with plants and trees, plant genomics, bioreactors for discovery, cell biology, process development in plant cell culture; and molecular marker technology and *in vitro* technology.
(b) *Animal biotechnology*: which includes animal nutrition and production, animal breeding and reproduction, animal health, diagnostics and biologics; and fish production and health.
(c) *Food biotechnology*: which includes process improvements to add value to commodity products, production and modification of foods and higher value ingredients; and development of methods of detection and safety evaluation of GM foods and bio-ingredients.

3.2.1   The nature and base of knowledge activities

Malaysia has a very good research base for the agriculture sector that can be traced back to its early colonial days and has since independence focused on the development of its plantation industry. This, in addition to the abundance of human resources for agricultural sciences is proving to be the main asset aiding its expansion into agricultural biotechnology. The developments in agro-based industries derive largely from availability of natural resources such as rubber and palm oil, which has grown to be a mainstay of the economy. The development of biotechnology is a natural progression from the earlier efforts in plantation industry and the need to create higher value in the agriculture sector.

Under the management of the Malaysian Ministry for Science, Technology and Innovation (MOSTI), biotechnology R&D activities in the country are categorized into seven sectors, namely, molecular biology, plant biotechnology, animal biotechnology, medical biotechnology, environmental and industrial biotechnology, biopharmacy, and food biotechnology. Present R&D activities in biotechnology are carried out mainly in research institutes and universities, while plantation firms and private laboratories concentrate mainly on TC. TC of several industrial crops (oil palm, rubber, rattan, forest trees) together with food crops (rice, banana, sago, herbs, and medical plants) and ornamentals (orchids and pitcher plants) has been successfully carried out in the country. All the R&D activities for each of the seven sectors fall under the Biotechnology Cooperative Centre (BCC) established to coordinate activities, which is overseen by a coordinator.

The development of biosciences within universities and public research institutes focuses on the understanding of the genetic make-up of crops such as rice, and the underlying mechanisms of disease and pest-resistance of crops. New varieties of horticultural crop such as orchids are being developed to obtain new hybrids in response to market requirements. Research in transgenic plant's that have new characteristics have been carried out in order to evaluate the potential of plant biopharming. Preliminary genetic maps for oil palm and rubber have been generated in efforts to increase the efficiency of conventional breeding. Several animal recombinant vaccines have been produced to assist the development of animal husbandry. Marker-assisted breeding strategies are also being practised to increase the efficiency of livestock breeding and to reduce imports of animal feed.

The application of bioremediation techniques in the treatment of industrial and agricultural wastes through bio-augmentation is already underway, while several diagnostics kits for dengue and other infectious tropical diseases have been produced. A major international partnership has been established with the Massachusetts Institute of Technology (MIT) under the Malaysian-MIT Biotechnology Partnership Programme (MMBPP). The programme focuses on natural products discovery and oil palm biotechnology, and has generated intellectual property rights (IPRs) over two natural products since its inception in 1999.

3.2.2 The main actors

The main actors in its agricultural biotechnology sectoral system are the government, PRIs, and university centres of excellence, Biotechnology

Coordination Centres (BCCs), and firms. The government is a very prominent actor, funding most of the on-going research activities and providing the framework for biotechnology development in the country. The extensive role of the government in Malaysian agricultural biotechnology is discussed in the section 3.5. Table 3.1 below contains a summary of the main research activities on-going in the universities and research institutions such as MARDI, UPM, UKM, USM, MINT, SIRIM and Technology Park Malaysia (TPM).

Table 3.1   Research institutions and universities engaged in biotechnology work

Institutions/ programmes	Focus area
Malaysian Agricultural Research and Development Institute (MARDI)	• Improved rice varieties • Disease resistance in rice, chilli, papaya • Delayed ripening in papaya • Floral colour and senescence in orchids • Advanced livestock breeding
Malaysian Institute for Nuclear Technology for Research (MINT)	• New plant varieties with improved characteristics • Production of phytocompounds through bioreactors • Microbial process for bioremediation and agro-waste treatments • Monoclonal antibodies for cancer • Blood products as raw material for drugs • Drug discovery for anti-cancer and anti-diabetics
Malaysia-MIT Biotechnology Partnership Programme (MMBPP)	• Natural products discovery • Oil palm biotechnology
Malaysian Palm Oil Board (MPOB)	• Yield improvement • Improve oil quality • Production of bio-plastics • Bio-fuel
Malaysian Rubber Board (MRB)	• Yield improvement • Disease resistance • Production of high value proteins
Technology Park Malaysia Corporation Sdn Bhd (TPM)	• Production of food and feed • Animal and aquaculture feed • Production of anti-cancer compounds from periwinkle • Production of nutraceuticals from micro-organisms • Development of biodiversity research and production centres

(Continued)

Table 3.1 (Continued)

Institutions/ programmes	Focus area
Universiti Kebangsaan Malaysia (UKM)	• Molecular biology of various pathogens causing tropical diseases • Antibody engineering • Molecular systematic studies of wildlife and domestic animals
Universiti Putra Malaysia (UPM)	• Oil palm expressed sequenced tags (ESTs) • Floral/meristem/embryo development • Plant defence stress response • Marine biodiversity • Diagnostic kit for aquatic animal disease • Bioremediation • Antibiotics and resistance genes
SIRIM Berhad	• Development of herbal-based cosmeceutical products formulation and manufacturing process • Production of industrial enzymes and fine chemicals through microbial fermentation • Development of environmental processes such as bioremediation and recycling technology • Providing environmental toxicity assessment and evolution services • Development of biosensor and sampling devices • Development of biotechnology measurement standards

Source: Compiled by authors, 2007 survey.

The seven BCC's created to coordinate the R&D activities for the biotechnology sectors play a very important liaison role. There are three BCCs of relevance to agricultural biotechnology and these are listed in Table 3.2 below.

Firms, both national and international, form the last category of important actors in the system.

3.3 Science, technology, and innovation investments

One of the distinguishing factors that separate advancing latecomers from latecomers is the recognition of, and the competence of, the former to leverage a wide variety of financing mechanisms to develop every aspect of the innovation continuum. The Malaysian government is the main financer of R&D for the sector and several initiatives exist for financing and commercialization of agricultural biotechnology.

Table 3.2 BCCs of relevance to agriculture biotechnology

BCCs	Coordinating institution	Projects
Animal Biotechnology Cooperative Centre	UPM	• Development of novel therapeutic and diagnostic reagents from Newcastle disease virus • Improvement of shrimp health through bioremediation and natural health food • Microbial fermentation of palm kernel cake for poultry feed • Development of a biotechnology for the production of single-sex cattle population based on *in vitro* embryo production, sperm separation, and embryo transfer • Development of recombinant bacteria for delivery of functional proteins • Alternative control of haemorrhagic septicaemia in beef cattle • Development of DNA test for marker-assisted selection in beef cattle • Application of a molecular biotechnology to breeding and management of Sea Bass (*Lates calcarifer*) • Investigation of the basic molecular biology properties of nipah virus • Genome mapping of Malaysian freshwater catfish • Determining and improving quality of shrimp eggs • Development of diagnostic method for rapid detection and differentiation of influenza virus subtypes
Food Biotechnology Cooperative Centre	UPM	• Production and application of enzymes for enhancement of sago starch • Improving the functionality and nutritional qualities of fats and oils using biotechnology processes • Production of microbial polyunsaturated fatty acids (PUFA) • Utilization of micro-organisms and their metabolites for the improvement of the fermentation processes selected Malaysian fish and vegetable/fruit-based foods

(*Continued*)

Table 3.2 (Continued)

BCCs	Coordinating institution	Projects
Plant Biotechnology Cooperative Centre	MARDI	• Breaking the yield ceiling of rice through gene/genome manipulation • Development of plant bioreactor technology for the production of fragrance compounds from *Michelia alba (Cempaka Putih)* • Improvement of commercially important orchids using biotechnology • Genetic engineering of mangosteen (*Garcia mansostana L*) for early flowering

Source: Authors' survey (2007).

3.3.1 Dedicated biotechnology policies for accelerated growth of the sector

One of such initiatives taken is Malaysia under the Commercialization Phase is the establishment of the Technology Acquisition Fund (TAF) Scheme.

In order to promote new knowledge and equally solve current socio-economic problems, the government has initiated a comprehensive funding mechanism covering the entire innovation spectrum to develop a sustained biotechnology industry. There is a clear consensus in the policy framework that following the developments in biotechnology in the frontier countries, whereby success depends in strengthening links between basic life science and applied research, the development and commercialization of biotechnology related R&D will depend to a great extent on building specific scientific infrastructural capacity, which evidently requires sustained financial investment. The TechnoFund is intended to provide the required funds for the development of technology commercialization within which biotechnology research is a prime area of focus. The fund finances research-based market needs while basic R&D for knowledge generation will continue to be funded under the ScienceFund.

It is estimated that the total amount of public and private investment under the National Biotechnology Policy over the period of 2006 to 2020 will reach a minimum of RM30 billion (exchange US\$1=RM3.50), of which medical and agricultural biotechnology will receive a substantial share, figure 3.1. This RM30 billion is expected to be funded by both the public and private sector, including, *inter alia:*

• government ministries and agencies;
• venture capital and debt venture communities;

Funding biotechnology

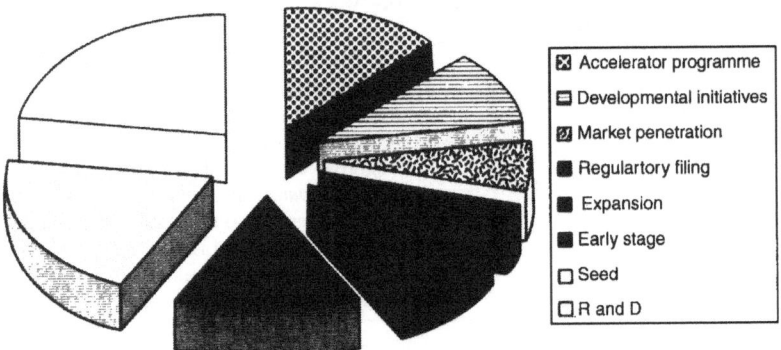

Figure 3.1   Funding allocation to biotechnology in Malaysia
Source: Biotechcorp, 2006.

- capital markets,
- financial institutions and institutional investors, and
- corporations including plantation companies, pharmaceutical companies, petrochemical companies, utilities, chemical companies, and government-linked companies (GLCs).

This total investment of RM30 billion made up of RM12 billion from the government and RM18 billion from the private sector is relatively minimal considering the total cumulative output expected from direct and indirect sources estimated at RM625 billion. This investment has the potential for creating more than 280,000 jobs or almost 2% of the country's job market. The output of the above investments is expected to be substantial. For instance, in 2004, biotechnology contributed less than RM100 million; in 2005 it is expected to record RM19.1 billion (representing 2.5% of nominal GDP); by 2010, RM50.8 billion (representing 4% of the nominal GDP); by 2015 RM100.9 billion (representing 5% of the nominal GDP) and by 2020 (see Figure 3.2). This represents a compounded annual growth rate (CAGR) of 23.7% per annum (Biotechcorp, 2007). The financial infrastructure is conceptualized on the premise of continual value addition with both the objective of solving socio-economic problems and wealth creation.

In addition, the government has introduced the Biotechnology Business Accelerator Programme (BBAP); a transaction-based strategy

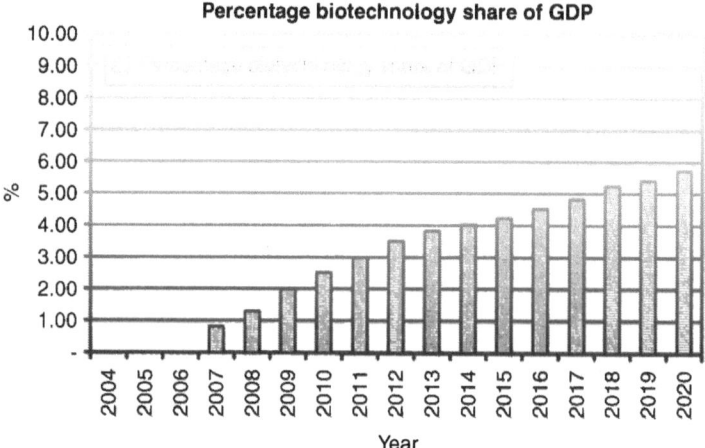

Figure 3.2 Percentage biotechnology share of GDP (projected)
Source: Computed by authors based on MOSTI's forecast, 2006.

that will be selectively applied to focus areas in agricultural, healthcare, and industrial biotechnology (Figure 3.3). The programme is expected to build upon Malaysian competitive advantages or areas where competitive advantages can be built quickly with each having unique key elements for success. The BBAP is presently being implemented and its first phase of execution is expected to show results by 2010.

The BBAP promotes four strategic approaches for the three sectors of biotechnology that are its key focus, namely, mentoring of research initiatives, establishment of centres of excellence, hub expansion and development of clinical expertise.

Clustering of existing knowledge bases in biotechnology in order to leverage on existing initiatives is underway through the initiative BioNexus Malaysia clusters. BioNexus Malaysia (see Box 3.1. below) is a group of specialized firms and institutions that support each other to create centres of excellence. Because of their inter-relationships, firms and organizations within clusters can take advantage of synergies in physical assets, human resources, entrepreneurship, and the sharing of ideas, which allows each firm and organization to perform at a global level what none would be able to achieve in isolation. The BioNexus Malaysia clusters are, in turn, linked to other regions across the globe with complementary specializations to complete specific value chains and to create new sources of technological innovation and wealth.

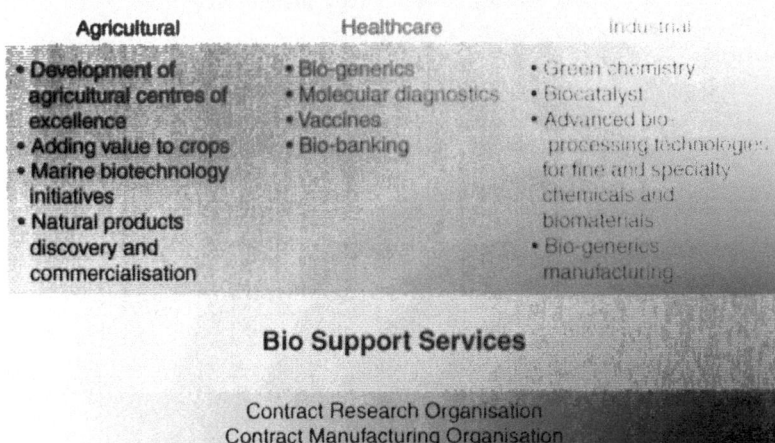

Agricultural	Healthcare	Industrial
• Development of agricultural centres of excellence • Adding value to crops • Marine biotechnology initiatives • Natural products discovery and commercialisation	• Bio-generics • Molecular diagnostics • Vaccines • Bio-banking	• Green chemistry • Biocatalyst • Advanced bio-processing technologies for fine and specialty chemicals and biomaterials • Bio-generics manufacturing

Bio Support Services

Contract Research Organisation
Contract Manufacturing Organisation
Biotechnology Equipment Manufacturing

Figure 3.3   Business accelerator programmes in key focus areas
Source: Adopted from National Biotechnology Policy of 2005.

Box 3.1 Clustering biotechnology initiatives: Bionexus Malaysia

BioNexus Status is a designation awarded to qualifying biotechnology companies, making them eligible for privileges contained within the BioNexus Bill of Guarantees. The Bill of Guarantees contains: freedom of ownership, freedom to source funds globally, freedom to bring required human skills from other countries into Malaysia, eligibility for competitive incentives and other assistance from the government, eligibility to receive assistance for international accreditations and standards, strong intellectual property incentives, access to supportive information network linking research centres of excellence, access to shared laboratories and other related facilities, and the change to access BiotechCorp as the one-stop agency.

BioNexus status companies may apply for the following tax incentives from the Malaysian government: (1) 100% income tax exemption for ten years commencing from the first year the company derives profit or Investment Tax Allowance of 100% on the qualifying capital expenditure incurred within a period of five years, (2) tax exemptions on dividends distributed by a BioNexus Company,

(3) exemption of import duty and sales tax on raw materials/ components and machinery/equipment, (4) double deduction on expenditure incurred for R&D, (5) double deductions on expenditure incurred for the promotion of exports, (6) a company that invests in its subsidiary, which is a BioNexus status company, is granted tax deduction equivalent to the amount of investment made in that subsidiary provided that the investing company owns at least 70% of that subsidiary, (7) a BioNexus company is given concessionary tax rate of 20% on income from qualifying activities for ten years upon the expiry of the tax exemption period, (8) a company or an individual investing in a BioNexus company is given a tax deduction equivalent to a total investment made in seed capital and early stage financing, (9) a BioNexus company undertaking a merger and acquisition (M&A) with a biotechnology company is given exemption of stamp duty and real property gain tax within a period of five years until 31 December 2011, and (10) a building used solely for the purpose of biotechnology research activities is given an Industrial Building Allowance over a period of ten years.

Source: Authors's survey (2007).

The central hub of the Bionexus is part of a greater nationwide cluster, in order to link biotechnology with other high technologies and eventually enable the growth of inter-disciplinary competencies such as bioinformatics. The central hub contains three institutes, each of which has a mandate to pursue research in the fields of technology that are critical to the biotechnology industry: genomics and molecular biology, nutraceuticals and pharmaceuticals, and agricultural biotechnology.

3.3.2 Persistence of R&D expenditures and governmental incentives

Our survey data confirms these investments and we measured through employment, percentage of gross domestic inputs, R&D expenditures, and R&D personnel that are available for 76 firms that we surveyed in each year of the period 2001–5 of the sample. In order to understand the kinds of activities on-going in the private sector, we performed a panel data analysis that investigates the dynamics of R&D expenditures. Table 3.3 and Figures 3.4–3.6 show the evolution of these variables over the whole period, while Table 3.4 shows maximum likelihood (ML) estimation results of the dynamic panel data type 1 tobit model that studies the persistence of R&D expenditures.

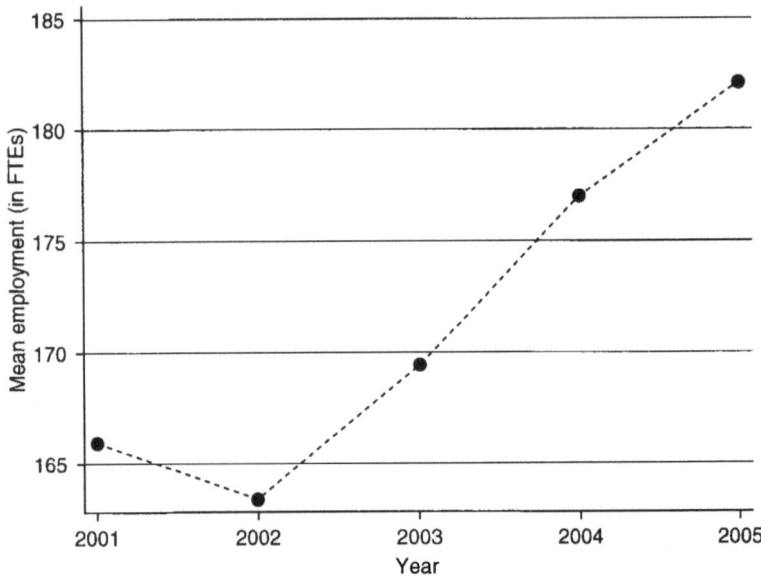

Figure 3.4   Mean employment over the period 2001–5
Source: Authors' survey (2007).

Figure 3.4 shows a decrease in firm mean employment between 2001 and 2002, and a steep increase from 2002 until 2005 as a response of the private sector to enhanced governmental support and initiatives for the sector.

Figure 3.5 shows an increase of the percentage of gross domestic inputs over the whole period.

Figure 3.6 shows a flat increase in mean R&D expenditures and R&D staff between 2001 and 2002, and a rather steep increase from 2002 until the end of the period.

The patterns of Figures 3.4–3.6 are also summarized in Table 3.3.

Table 3.4 below presents 12-point Gauss-Hermite ML estimation results of the dynamic type 1 tobit model. We consider two specifications of the individual effects. In the first pair of columns we consider the Wooldridge (2005) specification also known as the Chamberlain (1980) specification, and in the second pair of columns Mundlak's (1978) specification is considered. The results suggest that, regardless of the specification of the individual effects, there is evidence of 'true' persistence in R&D expenditures, i.e., after controlling for individual effects and modelling the correlation between the individual effects and the initial conditions, past R&D expenditures affect (strongly) positively and significantly current R&D expenditures. This is true in the Mundlak specification, and is insignificant in the Chamberlain

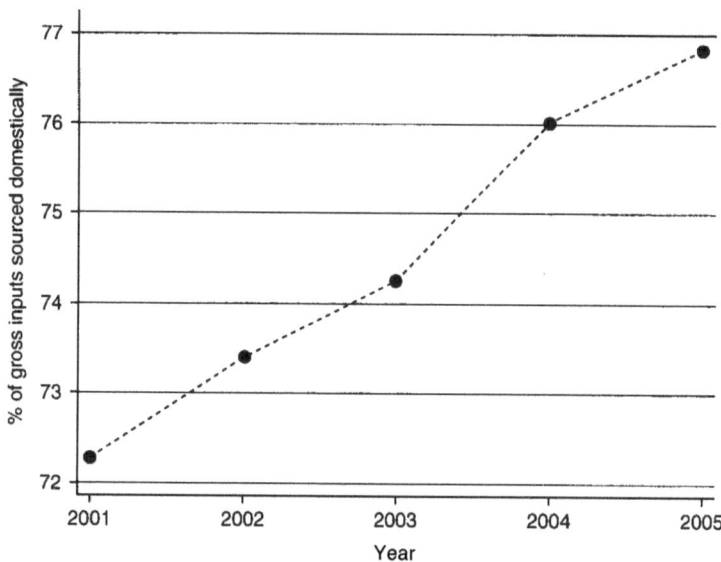

Figure 3.5 Mean domestic inputs over the period 2001–5
Source: Authors' survey (2007).

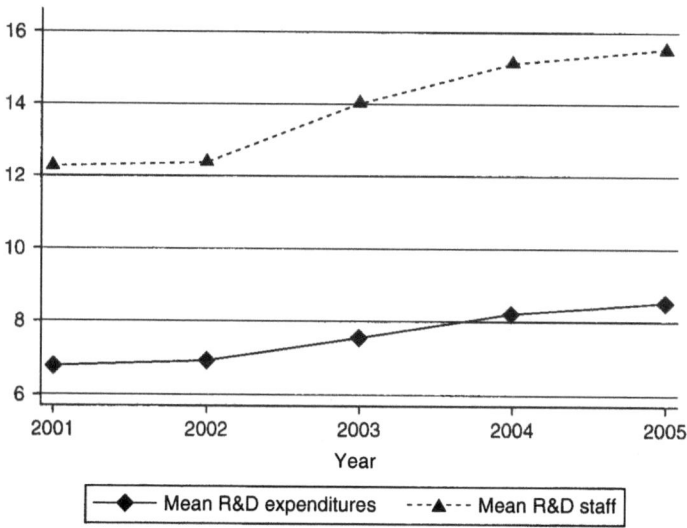

Figure 3.6 Mean R&D expenditures and R&D personnel over the period 2001–5
Source: Authors' survey (2007).

Table 3.3 Descriptive statistics of employment, domestic inputs, and R&D over the period 2001–5

Variable	Mean	(Std. dev.)	Min.	Max.
2001				
Total employment (FTEs)	166.000	(237.020)	3	1800
% of gross inputs sourced domestically	72.289	(30.291)	0	100
R&D expenditures (% of total workforce in FTEs)	12.263	(16.902)	0	50
R&D staff (% of total workforce in FTEs)	12.263	(16.902)	0	60
2002				
Total employment (FTEs)	163.447	(229.012)	3	1800
% of gross inputs sourced domestically	73.408	(28.498)	0	100
R&D expenditures (% of total sales)	66.901	(11.827)	0	50
R&D staff (% of total workforce in FTEs)	12.382	(16.797)	0	60
2003				
Total employment (FTEs)	169.447	(230.333)	3	1800
% of gross inputs sourced domestically	74.263	(27.150)	0	100
R&D expenditures (% of total sales)	7.559	(12.337)	0	60
R&D staff (% of total workforce in FTEs)	14.026	(17.253)	0	60
2004				
Total employment (FTEs)	176.987	(233.517)	7	1800
% of gross inputs sourced domestically	76.039	(26.260)	0	100
R&D expenditures (% of total sales)	8.217	(12.390)	0	60
R&D staff (% of total workforce in FTEs)	15.151	(17.790)	0	60
2005				
Total employment (FTEs)	181.961	(230.856)	5	1770
% of gross inputs sourced domestically	76.829	(25.481)	0	100
R&D expenditures (% of total sales)	8.545	(12.489)	0	60
R&D staff (% of total workforce in FTEs)	15.572	(18.289)	0	70
Number of firms		76		

Source: Authors' survey (2007).

specification. Equally, unobserved heterogeneity (through individual effects) plays a significant role in R&D expenditures.

3.4 Interactive learning and collaboration

Malaysia has made significant progress in the development of biotechnology especially since 2000, when biotechnology was prioritized as a lead sector. However, to enjoy a significant growth in biotechnology

Table 3.4 ML estimates of R&D expenditures

Variable	Coefficient	(Std. err.)	Coefficient	(Std. err.)
Current R&D expenditures (% of total sales)				
Past R&D expenditures (% of total sales)	0.881**	(0.087)	0.848**	(0.084)
Current total employment (FTEs)	1.160	(0.769)	2.147**	(0.084)
Intercept	−0.880	0.567	0.247	0.617
Extra parameters				
Initial R&D expenditures	0.138	(0.089)	0.165†	(0.086)
Total employment in 2002	−0.916	(1.032)	–	–
Total employment in 2003	−3.897*	(1.883)	–	–
Total employment in 2004	1.195	(1.484)	–	–
Total employment in 2005	2.611	(1.824)	–	–
Mean total employment	–	–	2.151**	(0.693)
σ_η	0.587**	(0.206)	0.889**	(0.191)
σ_ε	1.408**	(0.078)	1.405**	(0.078)
Number of observations		304		
Log-likelihood		−510.415–523.857		
Significance levels † : 10% * : 5% ** : 1%				

Source: Authors' survey (2007).

industry, greater capabilities are necessary to capitalize on the country's assets. A review of our survey data and the field interviews points to several underlying causes that need to be addressed to develop a competitive sector for agricultural biotechnology.

3.4.1 Lack of adequate human capital

Malaysia has always maintained a highly skilled and motivated workforce as a prerequisite for foreign companies investing in Malaysia, but also to enable the growth and competitiveness of local start-ups and small and medium enterprises (SMEs) within the country. For agricultural biotechnology, however, Malaysia needs a critical mass of high quality researchers and affiliated professionals. According to the MASTIC 2002 R&D Survey, the majority of biotechnology scientist and engineers (a total of 720 R&D personnel) are employed in the government research institutes and universities, whereas industry employed only 29 (4%) of the total. Approximately 100 PhDs in biotechnology research are in government research institutes, and approximately 150 are at the universities while less than ten serve the industry.

In the years 2001, 2002, and 2003 that the survey captured, a total of 370 PhDs graduated in all disciplines from Malaysian universities. It is estimated that 20% of these graduates are in disciplines relevant to biotechnology providing a pool of about 70 new PhDs per year in biotechnology. On the other hand, forecasting by AT Kearney (2003) estimates the demand to be 350 PhDs in five years in the health sector alone, therefore at least doubling of the pool is required to meet the manpower requirements of health biotechnology activities planned in the country. In addition it is estimated that 80 to 100 additional PhDs will be required for agricultural and industrial biotechnology. Over a 15-year perspective from 2005, it is estimated that 2000 scientists with PhDs will be required to support the R&D activities of 100-odd companies engaged in biotechnology by the year 2020. Apart from PhDs, the Malaysian government forecasts that a workforce of at least 10,000 skilled personnel in various aspects of biotechnology research (including marketing, organization, and legal skills) will be required to support the projected aims of the industry. This will translate into 280,000 additional jobs supporting the industry by the year 2020.

From 2006 onwards, a National Science Fellowship programme and other training programmes have been put in place to close this gap in human skills requirements, and it is expected that 5000 R&D scientist and skilled workers will be trained by 2010.

The government is also actively engaged in promoting collaboration between various Malaysian research institutes, local universities, and international research organizations to establish partnerships similar to the one between several Malaysian research institutes, local universities and the Massachusetts Institute of Technology (MIT) to explore natural product discovery and oil palm biotechnology, already discussed earlier in the chapter.

3.4.2   Collaboration and commercialization of R&D

The survey data sheds light on the extent and depth of systemic integration in the sector, and we present a descriptive analysis on the determinants of collaboration here. Collaboration with 12 types of organization active in the sectoral system of innovation is examined including PRIs, universities, industry associations, external/ private laboratories, hospitals, traditional knowledge holders, plant organizations, farmers groups, NGOs, GM companies and organizations, environmental organizations, and governmental agencies. Table 3.5 shows a correlation matrix with all these interactions and the main pattern of the table is that, with the exception of a few instances, collaboration among different types of organization is significantly interlinked. For instance, collaboration with PRIs is positively and significantly correlated with collaboration with industry associations.

We examined further the intensity of collaboration; that is, for each type of collaboration, the percentage of firms that deem that particular type of collaboration 'fairly strong', 'strong' or 'very strong'. For instance, 50% (38/76) of the firms rate collaboration with PRIs fairly strong, strong, or very strong. Furthermore, we considered differences and similarities across collaborators and non-collaborators. On this issue, the key pattern is that collaborators tend to be innovators. For instance, the percentage of firms that are involved in new product development is significantly (at the 5% level) larger in the population of firms that collaborate strongly with PRIs than in the population of non-collaborators. The same result holds for new process development with regards to the same type of collaboration. Again, firms that collaborate with PRIs spend on average significantly more on R&D and have significantly larger R&D personnel in 2005 than non-collaborators. These two types of firm do not significantly differ in terms of employment. In other words technological dynamism is associated with firms and organizations with a high propensity to engage in inventive activities and vice versa.

3.4.3 Institutions, government incentives, and innovation

Given the considerable financial and organizational investments being made by the Malaysian government, particularly in universities and PRIs, our survey sought to ascertain the interim impact of these efforts. Again the proxy that we used was the propensity to commercialize inventive activity through innovation.

Table 3.6 suggests that with the exception of local universities (for R&D collaboration) and venture capital availability that play a positive and significant (at the 5% level) role in new process development, the institutional variables are generally insignificant (at the 5% level) in new product and new process development. This is also confirmed through other

Table 3.5   Correlation between the types of collaboration

	PRIs	Industry assoc.	Univ.	Private lab.	Hospitals	Trad. med. pract.	Plant org.	Farmer groups	NGOs	GMOs org.	Envir. org.	Gvt. agencies
PRIs	1.000											
Industry assoc.	0.476**	1.000										
Universities	0.586**	0.392**	1.000									
External/private lab	0.487**	0.444**	0.2717†	1.000								
Hospitals	0.475**	0.389**	0.2567†	0.636**	1.000							
Trad. med. pract.	0.579**	0.423**	0.533**	0.430**	0.476**	1.000						
Plant organizations	0.546**	0.465**	0.541**	0.566**	0.452**	0.655**	1.000					
Farmer groups	0.476**	0.497**	0.427**	0.575**	0.523**	0.588**	0.685**	1.000				
NGOs	0.558**	0.493**	0.478**	0.463**	0.460**	0.611**	0.597**	0.743**	1.000			
GMO organizations	0.602**	0.502**	0.446**	0.688**	0.575**	0.544**	0.684**	0.758**	0.752**	1.000		
Envir. organizations	0.406**	0.593**	0.251*	0.653**	0.603**	0.464**	0.535**	0.726**	0.553**	0.716**	1.000	
Gvt. agencies	-0.084	0.2357†	-0.2567†	0.2783*	0.133	-0.140	0.070	0.133	-0.065	0.035	0.262*	1.000

Significance levels †: 10% *: 5% **: 1%

Source: Authors' survey (2007).

Table 3.6 The role of government incentives in new product and new process development: t- and z-tests

Variable mean	New product development		New process development	
	No	Yes	No	Yes
Govt. innov. Incentives	0.800	1.000	0.750	1.000
Scientific manpower	0.800	1.000	0.750	1.000
Local univ. for R&D	0.733	1.000	0.667*	1.000*
Local inst. for R&D	0.800	1.000	0.750	1.000
IPP	0.800	1.000	0.750	1.000
Infrast. services quality	0.800	1.000	0.750	1.000
Venture capital availability	0.733	1.000	0.667*	1.000*
Local SMI dev. schemes	0.733	0.857	0.667	0.900
Gvt.-firm-tech.transfer	0.800	1.000	0.750	1.000
Staff transf. to local org.	0.800	1.000	0.750	1.000
ICT services	0.800	1.000	0.750	1.000
Number of organizations	15	7	12	10

*These figures are on average statistically and significantly (at 5% level) different for innovating and non-innovating research institutes or universities on the basis of z-tests

Source: Authors' survey (2007).

econometric analysis that we carried out where new product development is shown to be significantly correlated with none of the institution variables, and new process development is shown to be positively and significantly (at 5% level) correlated with local universities (R&D collaboration) and venture capital availability. The two types of innovation are positively and significantly (at the 1% level) correlated, and the institution variables are shown to be related to each other. In sum the impact of state policies are still not directly felt in specific material instances to the extent they can be attributed to the present innovation propensity of these organizations.

3.5 The role of the state in promoting agricultural biotechnology

Malaysia has long supported domestic biotechnology industry for years, particularly in agricultural biotechnology R&D, and biotechnology was identified an issue of national interest as early as the 5th Malaysian Plan (1986–90). The National Biotechnology Directorate (BIOTEK) was established in 1995 to promote and coordinate biotechnology R&D activities in the country and to encourage private–public sector collaboration in the national biotechnology programmes. The government has also established a National Biotechnology and Bioinformatics Network (NBBnet) in 1998 to promote closer collaboration and networking within and outside the country.

However, biotechnology was prioritized as a potential lead sector only in the 8th Malaysia Plan of 2000–5. Despite its late start in prioritizing biotechnology as an area of technology focus, Malaysia is significant for its systematic state approach in promoting the sector.

Pursuant to its prioritization as a lead sector, the Malaysian government launched a 15-year National Biotechnology Policy (NBP) in April 2005 in order to provide the institutional base for the sector. The 8th Malaysia Plan 2000–5 (8MP) made the sector one of the five 'core technologies' identified as vital engines for growth towards making Malaysia a developed nation by 2020. Seven biotechnology-based sectors were also identified (see section 3.2) and biotechnology coordination centres (BCCs) were set up. Under the 9th Malaysia Plan 2006–10 (9MP), further dedicated biotechnology policies were enacted, especially those that cater to finance, risk allocation and technology acquisition, and incubation of relevance to the sector. Malaysian Biotechnology Corporation (Biotechcorp, already discussed in section 3.3) has been allocated RM100 million of commercialization funds comprising the Seed Fund, the R&D Matching Fund, and the International Business Development Matching Fund. With the aid of the National Biotechnology and Bioinformatics Network (NBBnet), databases and bio-information for local genetic resources and core R&D activities have been set up that facilitate the work of the BCCs. NBBnet has also facilitated the establishment of high computing facilities for protein modelling and DNA analysis.

The NBP represents a significant shift in focus and resource allocation, with the state emerging to be a key player. The NBP is a 15-year plan with an ambitious target of developing 20 world-class biotech companies by 2020. It has several core initiatives, including:

- Developing human capital through education and training. The plan intends to set up a fund to help pay for training and hiring skilled workers.
- Proving funding and support through the entire R&D cycle ('from lab to market'). This includes dedicated matching grants for biotechnology, including financial support in patent application.
- Initiating legal and regulatory changes, including increased incentives for commercialization through changes that gives researchers

ownership shares in intellectual property (IP), along with their employers and investors.
- Acquiring IP and technology to boost the local industry.
- Providing significant incentives to attract foreign and domestic investment to the industry.

The NBP is propose to be implemented in three phases, with specific goals (see summary in Table 3.7):

- *Phase I (2006–10)*: Build capabilities in advanced facilities, skilled human resources, and technology acquisition.
  Phase I will include adoption of policies, plans, and strategies by government. Henceforth, Advisory and Implementation Councils are established together with the establishment of Biotechcorp.
- *Phase II (2011–15)*: Channel science into technology, products, and spin-off companies.
  Phase II focuses on the theme of 'lab to market' whereby drug discovery and development and the creation of new products will be given greater emphasis. Strategies to acquire technologies through acquisition of firms and foreign direct investment (FDI) will be implemented.
- *Phase III (2016–20)*: Strengthen technology licensing and innovation.
  For phase III, in pursuit of establishing biotechnology industry as a global business, the plan is targeted to focus on consolidating the strength and capabilities in technology development and further develop expertise and strength in drug discovery and development, see Table 4.1.

3.6 Summing up

Although Malaysia's investments and prioritization of agricultural biotechnology have been relatively recent (the first phase of the national biotechnology policy begun only in 2006), Malaysia's nascent capacities in agricultural biotechnology stand out due to two major factors: its national innovation system is advanced and well coordinated, thereby enabling the firms and organizations to embark upon integrating the new technology into existing R&D bases relatively easily, and it has a very strong research and innovation base for agricultural innovation at both the firm and the farm level, as demonstrated by its success in oil palm technologies. Our survey captures the years 2000–5, wherein,

Table 3.7   Execution and implementation strategies

Phase 1 (2005–10): Capacity-building	Phase II (2011–15): Lab to market	Phase III (2016–20): Global business
• Adoption of policies plans and strategies • Establishment of advisory and implementation councils • Capacity-building in R&D • Industrial technology development • Develop agricultural, healthcare and industrial biotechnologies • Develop legal and IP framework • Incentives • Business and corporate development through accelerator programmes • Bioinformatics • Skills development • Job creation • Regional biotechnology hub • Develop Bio-Nexus Malaysia as a brand	• Develop expertise in drug discovery and development based on bio-diversity and natural resources • New products development • Technology acquisition • Promote FDI participation • Intensify spin-off companies • Strengthen local and global brands • Develop capability in technology licensing • Job creation	• Consolidate strengths and capabilities in technology development • Further develop expertise and strength in drug discovery and development • Leading edge technology business • Maintain leadership in innovation and technology licensing • Create greater value through global Malaysian companies • Re-branding of BioMalaysia as global hub

Source: National Biotechnology Policy, MOSTI (2006).

although the strategic focus had shifted to biotechnology, the first phase of the NBP had not yet been implemented and the dedicated biotechnology policies were just beginning to be put in place. This explains our survey results: mean R&D investments, R&D personnel and domestic inputs were slowly on the rise between 2000–5 as a result of the shift in policy focus, but at the same time, many the governmental initiatives that were in place then were not deemed to be very important by the firms for product and process innovation. This situation, it is expected, will gradually change, especially now that the government's dedicated policy instruments are gradually beginning to take roots.

4
Vietnam Biotechnology: Building Local Capacity

4.1 Introduction

The development of modern biotechnology in Vietnam can be dated to 1994, when according to a government Resolution 18/CP,[1] biotechnology was targeted as one of the country's sectoral priorities in terms of scientific research for the period of 1995–2010. According to this resolution, biotechnology is considered an essential prerequisite to achieve national goals for food, feed and fiber production, healthcare, and environmental protection.

In reality, however, biotechnology development in Vietnam focuses relatively more narrowly on the agricultural and forestry sectors, given the rich ecosystem in Vietnam. Research activities at government-affiliated institutes concentrate in the main on molecular biology and genetic engineering, microbial biotechnology for production of bioactive compounds, biofertilizers, environmental protection and soil remediation, enzyme biotechnology, plant biotechnology for *in vitro* conservation and propagation of superior germplasm (plant tissue culture for mass propagation of planting material for horticulture, fruit, and forestry industries; for developing high-yield plants). These research activities depend heavily on funding from the government and, so far, little commercial interest has been shown in this field, a characteristic shared by most latecomer countries whereby inventive activities do not get translated into profitable products. Despite an early start, due to a variety of institutional barriers and shortcomings in management, biotech development did not really take off in Vietnam until early 2000.

Internationally, Vietnam lags behind in the biotechnology sector by many decades; however, Vietnam is mastering modern biotechnology in selected areas of agricultural applications, and this chapter analyses the current strengths in its agricultural innovation system.

Although the Resolution 18/CP of 1994 marks a milestone policy initiative that triggered the development of biotechnology in Vietnam, relatively small financial investment was committed to biotechnology in the subsequent 20 years (approximately US$7.5 million only: MOST, 2003). For this reason the country did not succeed in building a dynamic innovation capacity for the biotechnology sector which would have attracted complementary investment and other critical actors to the sector.

In the context of the globalizing world, the sector faces a number of challenges such as: (i) the weak research and development capacity of research and development (R&D) organizations and enterprises; (ii) a lack of human resources and investment; (iii) a shortage of modern equipment and facilities and knowledge of international standard and regulations; (iv) the weak environment of competitive and innovative enterprises (MOST, 2003). At the same time, there are numerous opportunities available to both local and foreign actors to develop domestic modern technology as well as to gain entry to new and potential markets overseas.

In order to meet these challenges, the government has been paying more attention to supporting the sector. One of the government initiatives is a national programme for biotechnology development in the field of agriculture for the period of ten years, from 2006–15. This programme provides financial support for R&D organizations to conduct research and improve R&D capacities, build facilities, and train human resources. The budget for this programme, about US$63 million (from the state budget), supervised by the Ministry of Agriculture and Rural Development (MARD), is small by global standards but significant given Vietnam's relative poverty. There is a similar programme related to biotechnology R&D within the Ministry of Science and Technology but with a smaller budget also for the period of 2006–10.

These programmes together with the policy reform of public R&D organizations are expected to generate greater dynamism and foster the development of biotechnology capacity in Vietnam. The thrust of the public-sector reform is to transform previously publicly funded organizations into self-sustaining entities such that public R&D organizations will no more be dependent on the state budget for their operations including generating sufficient funds to pay staff salaries as well as operational costs, and in the process, address the needs of the market. However, these policies and programmes were only recently established and it is therefore too early to assess their impact on the development of biotechnology in Vietnam. It is not only the public R&D organizations that are expected to change to cope with requirements of market demand in the new context; the industrial sector is increasingly under pressure to address the home market and to exploit opportunities offered by new markets overseas. Enterprises will also have to pay more attention to and invest explicitly in learning about, global international standards and regulations, such as intellectual property rights (IPRs) regimes, good agricultural practice, and good manufacturing practices.

In sum, the development of biotechnology in Vietnam can so far be divided into the following periods: before 1994, from 1995–2000, from 2001–5, and after 2005. Before 1994, modern biotechnology was relatively poorly developed and mostly focused on traditional biotechnology. After Resolution 18/CP was promulgated in 1994, the period of 1995–2000 could be considered as the start-up phase of biotechnology development. The total expenditure for biotechnology R&D for this five-year period was US$2.5 million. This amount is equal to the total expenditure in biotechnology for the more than ten years from 1981 to 1995. However, the next period from 2000 to 2005 is a 'stagnation phase' in the development of biotechnology. In comparison with other sectors of industry and technology, the investment and efforts made in biotechnology development was meagre. The phase from 2005 to 2009 represents a revitalization phase. Understanding the importance of biotechnology for the social and economic development of the country, especially for an agriculture-driven country such as Vietnam, the state decided to invest more resources into developing this sector. A significant investment programme for the development of biotechnology in the field of agriculture with a total investment of US$65.2 million was set up to be monitored by the Ministry of Agriculture and Rural Development for the period 2006–15. The Ministry of Health and the Ministry of Science and Technology are also running other R&D programmes for the period 2006–10 relating to other fields of the biotechnology. In addition, in 2005, important laws and regulations related to technology development in general such as intellectual property rights and technology transfer were issued and are playing an important role in the revitalization of biotechnology in Vietnam. In other words, considerable organizational and institutional initiatives are ongoing in Vietnam. However, how much of this will impact on developing the capacity to innovate, is a subject to be discussed in subsequent sections.

4.2   The nature and base of knowledge activities

There are many players involved in the biotechnology innovation system. However, the system still lacks some critical intermediate actors and also the linkage between the actors is weak. Currently, the most active players in the system are R&D institutes, universities, and some enterprises.

Notably, most of the *R&D organizations* in Vietnam are public research organizations, whose funding depends on the state, while most of the agricultural R&D organizations are located in the Ministry of Agriculture and Rural Development (MARD). Private R&D organizations and private companies conducting R&D services are rare. In the field of biotechnology the number of R&D organizations in the agriculture sector is more than in pharmaceuticals and other sectors.

For building human capabilities, there are about 80 *universities* carrying out training activities related to agriculture, pharmaceuticals, and healthcare, while some of the R&D organizations also provide training courses – post-graduate and retraining. Also, there are some R&D centres/institutes under the universities categorized as R&D organizations as well, although, universities in Vietnam are more focused on the training function and less so on the research function. In the past, universities were mainly publicly owned, increasingly though, a number of non-public universities have emerged.

According to the statistics from Government Office of Statistics (GOS) (2005), the number of agricultural production and service enterprises in 2004 was 726, with food processing and beverage firms numbering 4484. The agricultural sector is sub-divided into sub-sectors namely: cultivation, fruit and vegetable, veterinary, sea products, plant protection, husbandry, meat processing, cashew nut processing, tea processing, and so on. Husbandry and food processing are the sub-sectors with the highest number of firms.

The previously large number of mostly state-owned enterprises in Vietnam, including those in agriculture is increasingly being equitized and evolving into shareholding companies, but the state still owns a large proportion of equity in the firms (usually 51%). Small and medium enterprises (SMEs) predominate in Vietnam and this is true for this sector as well; with SMEs accounting for more than 90% in total number of enterprises and growing, following after the policy of the government that encourages private ownership of businesses. However, most SMEs lack R&D capacity; they are short of resources to invest in R&D activities while intermediate organizations are equally unable to support them in promoting innovation in general (Nguyen Thanh Tung, 2004).

The intermediate agencies, largely *government agencies* such as the Ministry of Science and Technology, Ministry of Agriculture and Rural Development, are meant to promote innovation in firms, however, these actors themselves are limited both in providing finance capital as well as human resources (Nguyen Vo Hung, 2003). For this reason, their contributions to the innovation capacity of the sector is limited. We provide more detailed picture in the sector mapping that follows.

4.3 Science, technology, and innovation (STI) investments

In the period before 1994, because of low STI investment, the country's system of science had largely outdated equipments, poorly furnished laboratories, and unskilled research personnel; in other words, Vietnam had no access to modern scientific facilities and a limited pool of knowledge. Thus the bioscience and biotech system in Vietnam was largely non-existent while both state and non-state investment in biotechnology was episodic and fragmented with many projects uncoordinated across different set of actors. This inadequate investment and the suboptimal use of resources meant that the total investment from 1981 to 1993 for all bioscience- and biotech-related activities amounted to only US$10 million. The total expenditure for R&D activities in biotechnology was VND 80 billion (approx. US$5 million).

After 1994, especially between 2002 and 2007, Vietnam began to invest in modern technologies with emphasis on applied research focusing on local socio-economic problems. Following the resolution of the government to promote biotechnology development, some key national laboratories received funding between VND 3–9 billion. However, progress has been slow in developing the laboratories to effectively introduce new equipment in order to raise utilization capacity, as well as improve staffing. Due in part to established patterns of behaviour and difficult to unlearn social habits and practices, investment in biotechnological research and transfer has been slow in taking effect and the use of these resources has not been effective. Due to historical reasons, Vietnam has been slow in attracting investment in biotech R&D from other sources (Phan Van Chi, 2004).

In general, over the 1997–2007 period, a system of laboratories from central to local levels began to emerge with relatively modern equipment. However, operational expenditure has been insufficient because according to scientists, there is no institutional guidance on what are routine – as opposed to – strategic tasks and, equally, there is no meaningful link between theoretical, applied research, and practical production. Hence, on one hand, the use of newly acquired laboratories has not been properly defined particularly in light of the reform compelling laboratories to generate their own revenues, while, on another hand, biotechnology R&D has been carried out mainly within the national biotechnology programmes. Scientific effort has been driven largely by the professional background of the scientists with little consideration for market needs. For instance, a scientist with a background in enzyme technology would inevitably propose to develop research in this field and the laboratory will therefore develop expertise in enzyme research. As a result, though several outcomes of the research were deemed complete, not many of them have found practical use in production, because the overall national programme has not reconciled the scientific capabilities and the supply of skills with market needs.

The role of firms, especially private-sector enterprises in biotechnological research and application is still evolving and unclear in part because the driving force for development and competitive capability of the state-owned enterprises in biotechnology has also been weak. The system linkage among key actors in the biotechnological innovation system (e.g. enterprises, research institutes, and universities), is similarly evolving and poorly articulated. (Tran Ngoc Ca, 2003). The current relationship between R&D organizations and enterprises is constrained by a number of factors such as a shortage of funds and capable human resources within the firms themselves and collaboration with R&D organizations is still constrained by poorly defined rules of engagement. When cooperating with foreign firms, Vietnamese firms tend to follow the initiative of their partners; Vietnam still lacks the linkage and cooperation experience, as well as qualified research teams with international research capability to match their foreign counterparts.

The overall direction of training has many shortcomings: there is lack of a well-designed strategy and specific structure for meeting the capacity gap and scientists point to the fact that curricula have not been able to meet demand including post-graduate training.

The institutional device for signalling the direction of biotechnology development industry is still embryonic; the production base is relatively narrow, bound to the national market through the primary, unprocessed bio-related product demand, which consists mainly of raw materials such as tea, coffee, rubber, rice, and aquaculture. These kinds of product have enjoyed only temporary comparative advantage but have weak long-term competitiveness (Nguyen Manh Quan, 2004).

Overall, the system of innovation in general, not only the biotechnology innovation system is still quite weak, important actors are missing or do not play their role well in the system, sometimes even hampering the innovation process due to negative social norms and habits. The intermediate organizations such as those supporting technology transfers (technology transfer offices) are few. Most of R&D organizations and universities do not have technology transfer offices to deal with commercialization of their R&D results. And the most important of the system – the linkage between the actors – is extremely weak and needs to be improved.

Realizing these weaknesses, Vietnam has taken several steps to build up the innovation systems at national and sectoral levels. National R&D programmes, including biotechnology R&D programmes, are selected carefully by tender to ensure quality and conformity with the broader vision. A number of R&D organizations, especially in engineering, are now being transformed to operate independently (just like the firms), without funding from the state as the case was in the pre-market days. However, the policy still has many critics, especially the privatization of public R&D organizations in the agriculture sector, because poor farmers are their main customers and they risk exclusion from scientific knowledge and R&D results for which they often cannot pay. With these concerns, this policy is being implemented incrementally – to attenuate inevitable negative feedbacks that may result from institutional experimentation – through learning.

It is extremely important for Vietnam to develop and support smallholder farmers because agriculture contributes a significant proportion of national product and stabilizes the economy and society. Agricultural biotechnology application, therefore, is judged critical for increasing crop production to meet domestic and export market demands and as well, for conserving natural resources.

In the field of plant biotechnology, technologies of forming new breeds such as tissue culture and selecting cell generation have been mastered and applied widely to production of rice, cross-breeding rice, maize, and so on. Biotechnology has contributed to the achievement of self-sufficiency of 25% demand of F1 cross-breeding rice, thereby generating economic revenue of approximately US$25–43 million per year (MARD, 2003). New rice production has moved Vietnam to be the second largest rice exporter in the world. By building capability in applied genetic technology in forming plant varieties, Vietnam has completed the genetic transfer processes of anti-insects, anti-weeds, and anti-disease caused by fungus and bacterium. The mapping of the rice gene has also been accomplished. Multiplication, improvement, and disease elimination technology for key plants (eucalyptus, gum tree, orange tree, mandarin tree, sugar cane, banana, orchid, etc.) researched by local scientists have been applied nationwide and have become the significant technologies for agricultural improvement.

In the field of animal biotechnology, Vietnam has mastered artificial insemination, preservation technology, embryo transfer, and vaccine production technology. The veterinary industry has largely mastered vaccine production and has mostly met the demand of most animal species as well as quantity used in livestock farming using micro-organic fermentation technologies and animal cell technologies. Three vaccine and veterinary firms in Vietnam have met all the needs of husbandry and households. Also in the field of husbandry, agricultural firms implemented the mass production of frozen cow and ox sperm, and embryo worth approximately US$1.2 million in 2001–2 alone (MARD, 2003). Currently a number of research institutes have started to pay attention to animal cloning biotechnology.

In aquaculture, biotechnology has been applied in breeding and in preventing diseases. In terms of breeding, Vietnam has become the world's highest breeder of crabs, and is the second country, following Taiwan, in successfully producing garrupa – a kind of economically beneficial seafood. Recently, DNA technology in estimating early symptom in sprawl has successfully been applied to help farmers raise their income. Agriculture accounts for 22% of Vietnam's economic output and over two-thirds of employment, primarily based on small family farms (MARD, 2003). However, the sector faces inevitable difficulties and challenges particularly after joining the WTO. This has opened up significant opportunities for Vietnam agriculture's products to exploit new market. Reaping the benefits of new markets will help largely in solving problems of the innovation system, in order not to lose market share to other emergent competitors. Problems relate largely to limited use of human resources, low investment in R&D capacity, as well as disarticulation in the actors' linkage configuration.

4.4   Survey results: Employment, domestic inputs and R&D staff over the period 2001–5

This section assesses the impact of the initiatives taken so far, as well as the outcome of the reform, no matter how tentative. It is based on a questionnaire survey administered to several R&D organizations, universities, and firms in Vietnam in the third quarter of 2007. This was followed by extensive face-to-face interviews of selected organizations all of which informed much of the perspectives of this chapter. The data collected also provide information on the role of institutions and government incentives in promoting innovation, the dynamics of R&D investments, and the determinants of collaboration in the Vietnamese biotechnology sector.[2]

Some of the variables are not available in the survey for all the years of the period 2001–2005, which makes the use of dynamic panel data models impossible. The variables used are, among others, total employment measures in full-time equivalents (FTEs), the percentage of gross inputs sourced domestically, R&D intensity measured as the ratio of R&D expenditures over total sales and R&D staff as a percentage of total workforce also measured in FTEs.[3] We show how these variables change over the period 2001–5 in Table 4.1 and Figures 4.1–4.3 which describe and estimate a panel data model studying the persistence of R&D investments the estimation results of which are reported in Table 4.1.

Table 4.1 Descriptive statistics of employment, domestic inputs, and R&D over the period 2001–5

Variable	Mean	(Std. err.)	Min.	Max.
2001				
Total employment (FTEs)	286.682	(973.095)	1	7500
% of gross inputs sourced domestically	31.152	(40.097)	0	100
R&D expenditures (% of total sales)	2.783	(9.198)	0	65
R&D staff (% of total workforce in FTEs)	3.600	(11.765)	0	73
2002				
Total employment (FTEs)	297.848	(1032.119)	1	8000
% of gross inputs sourced domestically	33.977	(40.878)	0	100
R&D expenditures (% of total sales)	3.942	(11.084)	0	65
R&D staff (% of total workforce in FTEs)	3.381	(9.734)	0	50
2003				
Total employment (FTEs)	334.182	(1016.457)	1	8000
% of gross inputs sourced domestically	40.152	(41.604)	0	100
R&D expenditures (% of total sales)	4.509	(10.95)	0	65
R&D staff (% of total workforce in FTEs)	4.117	(12.087)	0	73
2004				
Total employment (FTEs)	451.864	(1004.819)	1	8000
% of gross inputs sourced domestically	50.758	(40.307)	0	100
R&D expenditures (% of total sales)	5.619	(11.188)	0	65
R&D staff (% of total workforce in FTEs)	4.626	(10.92)	0	53
2005				
Total employment (FTEs)	505.485	(1119.47)	1	9000
% of gross inputs sourced domestically	58.883	(36.466)	0	100
R&D expenditures (% of total sales)	5.765	(10.892)	0	70
R&D staff (% of total workforce in FTEs)	7.148	(13.119)	0	73
Number of firms	66			

Table 4.1 reports descriptive statistics of the above-mentioned variables for each year of the period 2001–5. With the exception of R&D staff that decreases from 2001 to 2002, there is a systematic increase in the mean of the above-mentioned variables over the whole period. For instance, total employment increases from about 287 FTEs in 2001 to 505 FTEs in 2005. The variation in the mean of the variables over time is also shown in Figures 4.1–4.3.

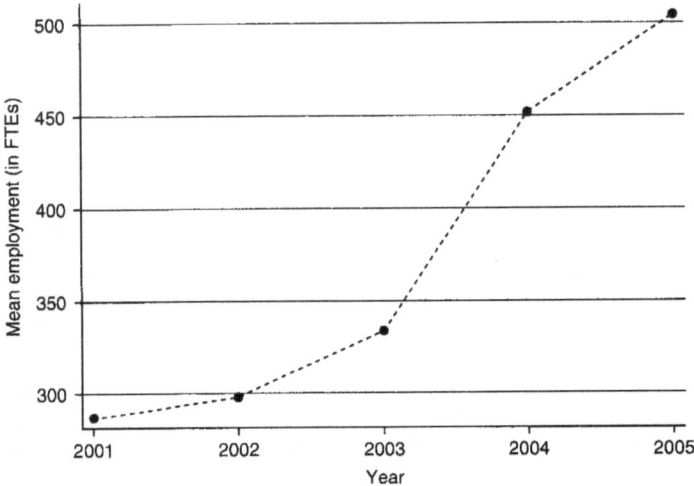

Figure 4.1   Mean employment over the period 2001–5

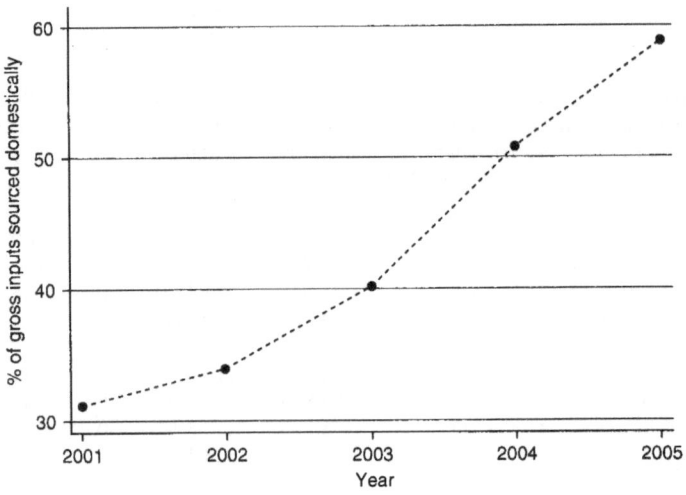

Figure 4.2   Mean domestic inputs over the period 2001–5

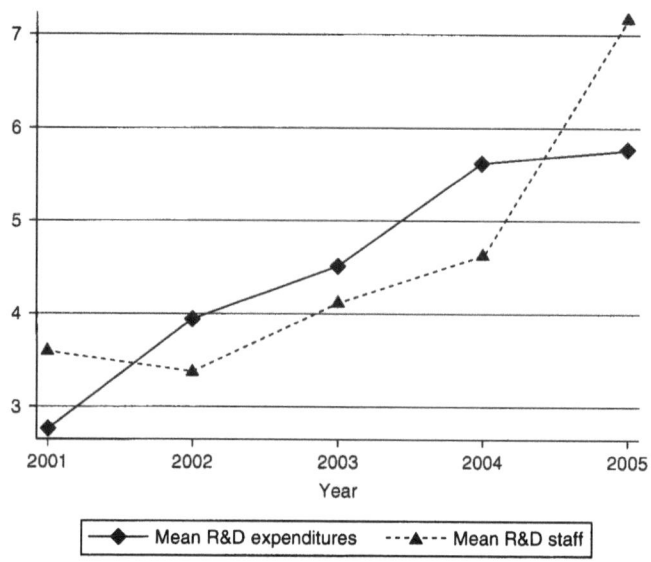

Figure 4.3 Mean R&D expenditures and R&D personnel over the period 2001–5

We now describe the model of the persistence of R&D investments (see Table 4.2).

The results suggest strong 'true' persistence in R&D investments, that is, after controlling for the individual effects and modelling the correlation between the individual effects and the initial conditions, past R&D investments affect strongly positively and significantly current R&D investments. Current employment also affects positively and significantly current R&D investments and unobserved heterogeneity plays a highly significant role in current R&D investments.

4.5 Key actors and organizations and their roles in the sector

As mentioned in section 4.3, the most active players of the system are R&D organizations, universities, and enterprises. R&D organizations in Vietnam traditionally play the role of 'knowledge creators', technology suppliers, and service providers. Universities provide human resources for other actors and also conduct some R&D activities. Firms are supposed to be the major players of the sector in the innovation system, however, most of them do not conduct R&D by themselves but depend on the R&D results of public organizations and/or out-source overseas. The intermediary actors are largely missing in the sector, especially

Table 4.2   ML estimates of the dynamic panel data[4]

Variable	Coefficient	(Std. err.)
Current R&D expenditures (% of total sales)		
Past R&D expenditures (% of total sales)	0.817**	(0.080)
Current total employment (FTEs)	1.293**	(0.317)
Intercept	−2.041	(2.180)
Extra parameters		
Initial R&D expenditures	0.241**	(0.088)
Total employment in 2002	−0.240	(0.312)
Total employment in 2003	−0.271	(0.342)
Total employment in 2004	0.266	(0.462)
Total employment in 2005	−1.043[†]	(0.565)
σ_η	1.421**	(0.430)
σ_ε	7.342**	(0.467)
Number of observation	264	
Log-likelihood	−562.511	

Significance levels: [†] : 10% * : 5% ** : 1%

technology transfer offices and relevant standard organizations. Most consulting firms provide financial and legal services rather than technology services while the government and its agencies also play an important motivational role in the sector. These agencies are not only important in providing finance for biotechnology development but also in creating a dynamic for biotechnology development through a number of related R&D programmes. However, the state's investment in the the sector is still not considered adequate relative to the sector's importance to the country's socio-economic wellbeing (MOST, 2003). In what follows we provide more details on the major actors of the system.

4.5.1   R&D organizations: Biotechnology laboratories

Presently (2008), MOST provides finance to establish and improve equipments through a network of about 60 biotechnology laboratories, located in several parts of the country. Half of these are TC laboratories to foster the production of crop plants, funguses, medicinal funguses, micro-organism fertilizer, biological pesticide, and so on. The average capacity of each laboratory is 0.5–4 million plants per year.

In order to advance biotechnology development in Vietnam, in 2002 the government established six key national biotechnology laboratories, and recently invested in supplementary laboratories for TC technology in South Vietnam. So far, six national biotechnology laboratories as well as 60 other laboratories are important sources for improving biotechnological research, training, and production in Vietnam.

Beside finance provided through the Ministry of Science and Technology, other sources such as ODA, have been mobilized to improve infrastructure facilities, such as laboratories, equipment, and quality control units, within the Ministry of Health, Ministry of Agriculture and Rural Development, the Ministry of Aquaculture, the Ministry of Defense, the Ministry of Industry, the Ministry of Resources and Environment, and university laboratories.[5]

In general, over the ten years, from 1995 to 2006 a network of laboratories, which were in poor condition, has been improved and furnished with relatively modern equipment to support notable projects in biotechnology. However, problems related to a lack of human resources remain to master and effectively use these modern equipments as well as lack of institutional mechanism to increase the effectiveness of the laboratories such as sharing laboratories between R&D organizations and industry.

4.5.2 R&D institutes

As noted earlier, biotechnology research and development in Vietnam is mainly financed and undertaken by the public sector. This is driven by several ministries such as: The Ministry of Agriculture and Rural Development, the Ministry of Health, the Ministry of Fisheries, the Ministry of Industry, Ministry of Science and Technology, depending on whether it is for agriculture or other fields. The Vietnam Academy of Science and Technology and many other ministerial institutions and centres are major actors involved in biotechnology research and development.

The key institute for agricultural biotechnology research is the Institute of Biotechnology (IBT), which is part of the Vietnam Academy of Science and Technology. IBT is considered a leading biotechnology research institute in Vietnam, conducting research activities on five major fields of biotechnology: (i) genomics and proteomics – the application of molecular technologies for classification, characterization, and conservation of the specialty and diversity of genetic resources; identification and functional analysis of new genes for engineering purpose; providing nationwide online service of genome informatics; (ii) biotechnology of micro-organisms – selection, evaluation and exploitation of new microbial strains for use in agriculture, bio-pharmacy, food processing and waste treatment, development of microbial fermentation and platform technologies for the effective expression and economical production of recombinant proteins and bio-active compounds; (iii) enzyme biotechnology – the purification and protein engineering of commercial important enzymes, screening of target proteins of bio-pharmaceutical value, design and development of biologically active peptides; (iv) plant biotechnology – the development of plant cell biotechnology methods for conservation and propagation of crop plants and plant germplasms, modification of crop traits by means of clone cell selection and genetic engineering for creating adapted cultivars with better quality and stress tolerance; (v) animal biotechnology – the use of animal biotechnological methods for controlling reproduction of livestock (cryo-preservation of sperm, embryo transfer), development of transgenic animals, development of animal cell culture for diagnostic purposes, *in vitro* fertilization and cloning.

The Institute of Agricultural Genetics (IAG) under the Ministry of Agriculture and Rural Development is also one of the very active R&D organizations of the system. IAG research activities focus on plant genetics to select and create new varieties of rice, maize, and soybean with high yields, good quality, and resistance to biotic and a-biotic stresses. In general, their research focus on the application of modern genetic methods and biotechnologies for the selection and creation of new crop varieties with high yields, good quality, resistance to pests and diseases, and tolerance to environmental stresses. IAG also conducts research on the improvement of micro-organisms for the conservation and processing of food and foodstuffs, and to produce bio-products for sustainable agriculture and environmental protection.

The Mekong River Delta Rice Research Institute is the leading organization in rice research. The institute carries out research on other crops and agricultural systems of the delta region, including surveying, genetics and plant breeding, plant protection, agronomy, soil fertility management, water management, post-harvest technology, farm machinery, physiology and ecology, variety and cropping systems, economics.

In the production of vegetable and fruit, there are several important R&D institutes such as SOFRI (Southern Fruit Research Institute) and RIFAV (Research Institute for Fruits and Vegetables). These institutes carry out breeding and variety selection of high yield and high quality fruit and vegetable crops to be developed in different ecological regions for local consumption, processing, and export. They also engage in research to improve processing and post-harvest technology and provide consultancy services and training in vegetable and fruit technology.

Although the number of R&D institutes that conduct research relating to biotechnology has increased recently, these institutes tend to work independently of one another with the result that research activities are not always coordinated. This difficulty in proper coordination, planning, and management of government investment in all aspects (research facilities and personal training) results in duplication of research efforts, and waste. Moreover, a number of research projects do not address the need of the market in part because R&D institutes usually do not have an office or unit to carry out technology transfer activities. In sum therefore, the linkage between industry and R&D organizations is weak.

4.5.3 Universities

In universities, the system of biotechnology and bioscience focuses mainly on teaching, with less emphasis on research. The Ministry of Education and Training, other ministries and S&T organizations have various training programmes for BScs, MScs, and PhDs of biotechnology in several formats. Some 12 universities have established training programmes and centres of biotechnology (see Table 4.3 below).

Hanoi University of Technology is one of the universities providing training courses in biotechnology. The training courses provided focus mainly on chemical biotechnology such as physical chemistry technology, organic chemistry and petrochemical technology, technology of pharmaceutical chemistry and pesticides for agriculture, polymer

Table 4.3 Universities in Vietnam providing biotechnology programme

No. University	Starting year of biotechnology programme
1 Hanoi University of Natural Science (under Hanoi National University)	1997
2 Hanoi University of Technology	1997
3 University of Natural Science, Ho Chi Minh City (under National University of Ho Chi Minh City)	1999
4 University of Technology, HCM City (under National University of Ho Chi Minh City)	2000
5 University of Agriculture and Forestry, HCM City	2001
6 Can Tho University	2001
7 An Giang University	2002
8 Da Lat University	1998
9 Tay Nguyen University	2000
10 Van Lang University	1995
11 Open University of Hanoi	1997
12 Open University of HCM City	1991

Source: MOST (2003).

technology, cellulose and paper technology, technology of inorganic chemistry and fertilizers, chemical engineering, chemical and food engineering. The university also conducts research (by the faculties and institutes under the universities) on several issues such as upgrading the quality of products from traditional materials, the effective use of natural resources in Vietnam, synthesis of fertilizers, pesticides for agriculture, pharmaceutical products, food technology, and biotechnology.

Hanoi University of Natural Science is another large university providing training on biotechnology as well as carrying out research in this field. The training courses related to biotechnology are provided by the Faculty of Biology in areas of Invertebrate Zoology; Vertebrate Zoology; Botany; Microbiology; Anthropology and Physiology; Cytology, Histology, Embryology and Biophysics; Biochemistry and Plant Physiology; and Genetics. Biotechnology researches are conducted by this faculty and two major centres under the university: the Center for Molecular Biology and Cell Technology and Center for Biotechnology. The research areas are related to: the development of animal germ cell technology, the development of recombinant DNA technology for production of proteins/enzymes of medicinal and agricultural use, molecular characterization of oxidative stress, investigation and characterization of natural bio-active compounds, application of germ cell technology of domestic animals for development of biotechnology in husbandry and veterinary, development of plant and TC technology for improvement of some fruit and crop varieties, etc.

In Ho Chi Minh City, there are a number of universities that provide training in biotechnology such as the University of Natural Science, Agriculture and Forestry University, University of Technology. In different regions and provinces, a range of Agro-Forestry universities have faculties conducting research and training activities related to biotechnology. However, as with other universities in the country, the budget for research activities are usually small and insufficient (MOET, 2003).

In general, so far, training has been conducted for more than 4000 staff, of which 1500 obtained BA degrees, 400 MSc degrees, and 90 PhD degrees. But this number has fallen short of demand, especially in the major fields of biotechnology such as genetic technology, cell technology, enzyme-protein technology and micro-organism technology. Along with universities, R&D institutes have also participated in training biotechnological staff. During the five-year period from 1996 to 2000, 23 MScs and 22 PhDs were trained through this research programme.

Increasingly, overseas training and programmes of internships leading to MSc and PhD training have been carried out with state budget funds of VND 100 billion per year (or approximately US$7 million). Through the Ministry of Education and Training a number of scientists have been trained in biotech areas. In the framework of national research programmes on science and technology, hundreds of biotechnology staff have collaborated with colleagues from developed countries through visits or research projects. In addition, scientists from other advanced countries have been invited to train Vietnamese in research methodology with more than 200 groups of biotechnology scientists participating in recent years. Vietnam has also relied on some overseas scholarship schemes to send students to the US, Europe, Australia, and Japan for training. However, on graduating, students tend to find the right professional jobs abroad and thereby reduce the pool of skills that are readily available in Vietnam because modern biotechnology is not well developed and working facilities are limited (MOET, 2003).

In general, universities in Vietnam pay more attention to theoretical training rather than to practice, due in part to a lack of equipment and facilities for students, and also due to a lack of faculties for those who have trained in modern and advanced biotechnology.

4.6   Survey results: Institutions and incentives for innovation

In order to understand the contribution of institutions and government incentives to changes made to products and processes in biotechnology, we used the proxy 'propensity to develop new product and/or process'. Table 4.4 shows descriptive statistics of innovation and institution variables and reports maximum likelihood (ML) estimates of a univariate probit model of product and/or process innovation and marginal effects for the 'best' specification of the model.

Table 4.4 shows that about 70% of the firms in the sample are involved in new product development, 71% are so in new process development, and 77% are involved in new product and/or process development. New product development is highly positively (about 0.7) and significantly (at 1% level) correlated with new process development. The institution variables that explain product and/or process innovation are all binary with value one if their contribution is deemed fairly strong, strong or very strong, and zero otherwise. For instance, 86% of the firms deem the contribution of government incentives fairly strong, strong, or very strong.

Table 4.4   Institutions and incentives for innovation: Descriptive statistics

Variable	Mean	(Std. dev.)	Min.	Max.
Dependent variables				
New product development	0.697	(0.463)	0	1
New process development	0.712	(0.456)	0	1
New prod. and/or proc. dev.	0.773	(0.422)	0	1
Independent variables				
Gvt. innovation incentives	0.864	(0.346)	0	1
Scientific/skilled manpower	0.970	(0.173)	0	1
Local univ. for R&D collaboration	0.682	(0.469)	0	1
Local research inst. for R&D collaboration	0.712	(0.456)	0	1
Intellectual property protection	0.667	(0.475)	0	1
Quality of local infrast. services	0.864	(0.346)	0	1
Availability of venture capital	0.303	(0.463)	0	1
Local SMI develop. schemes	0.636	(0.485)	0	1
Gvt.-firm-tech. transfer. coord. councils	0.697	(0.463)	0	1
Trans. of staff to local firms or R&D inst.	0.712	(0.456)	0	1
Number of firms		66		

Sector Performance: Engaging in new product and process development

Ideally, the model should be a bivariate probit explaining new product development together with new process development where both types of innovation are allowed to be correlated. However, the data does not allow the estimation of such a model as it is not sufficiently informative. We then consider a univariate probit model explaining the likelihood of being involved in new product and/or process development.

Table 4.5 reports ML estimates of the model. Two specifications are considered, namely the unrestricted one where innovation is explained by all the institution variables and the restricted one where some coefficients are restricted to be zero. The joint significance of those coefficients is tested using a likelihood ratio test that clearly shows that the restricted model is not rejected (at 10% level), hence the preferred model. The results show that intellectual property protection (IPP), the quality of local infrastructure services and the participation in local SMI development schemes affect positively and significantly new product and/or process development while the availability of venture capital and the transfer of personnel to local firms or R&D institutions (for training) play a negative role in new product and/or process development.

Table 4.5 Probit estimation results and marginal effects: New product development

Variable	Coefficient	(Std. err.)	Coefficient	(Std. err.)	Slope	(Std. err.)
	New product and/or process development					
	Unrestricted model		Restricted model		Marginal effects	
Gvt. innovation incentives	0.180	(0.722)	–	–	–	–
Scientific/skilled manpower	1.015	(1.276)	–	–	–	–
Local univ. for R&D collabo.	−1.676	(1.968)	–	–	–	–
Local research inst. for R&D collaboration	1.755	(1.964)	–	–	–	–
Intellectual property protection	0.940†	(0.512)	1.099*	(0.486)	0.290*	(0.134)
Quality of local infrast. services	1.050†	(0.611)	1.131†	(0.591)	0.350†	(0.209)
Availability of venture capital	−0.931†	(0.563)	−1.090*	(0.491)	−0.295*	(0.145)
Local SMI develop. schemes	0.984	(0.614)	1.022*	(0.482)	0.260*	(0.132)
Gvt.-firm-tech. transfer. coord. councils	−0.004	(0.661)	–		–	
Trans. of staff to local firms or R&D institutes	−1.910*	(0.914)	−1.739*	(0.711)	−0.270**	(0.082)
Intercept	−0.613	(1.141)	0.291	(0.564)	–	
Number of firms			66			
Log-likelihood	−24.426		−25.501			

Significance levels: † : 10% * : 5% ** : 1%

4.7   Interactive learning in the biotechnology system

Again relying on the survey data, this section discusses one of the key elements of systems, the nature and intensity of systemic collaboration. We present a descriptive analysis of the determinants of collaboration. Twelve types of collaboration are considered in the study.[6] The collaboration variables are binary with the value one if the intensity of collaboration with the institutions is 'fairly strong', 'strong' or 'very strong', and zero otherwise. The appendix Table 4.7 shows the matrix of the 12 types of collaboration. It suggests, for instance, that collaboration with public research organizations is highly significantly (at 1% level) and positively correlated with collaboration with industry association, universities, external/private laboratories, and governmental agencies, and significantly (at 5% level) and positively correlated with collaboration with hospitals and non-governmental organizations (NGOs).

The table provides objective ranking for each type of collaboration, the percentage of firms for which that particular type of collaboration is fairly strong, strong, or very strong. For instance, 65% (43/66) of the firms have fairly strong, strong, or very strong collaboration with PRIs. We also tested for the characteristics of both collaborators and non-collaborators for each type of collaboration. The main pattern is that, regardless of the type of collaboration, collaborators and non-collaborators have similar characteristics in the sense that they are not statistically and significantly different on the basis of t- and z-tests. Some noteworthy exceptions concern collaboration with PRIs that is driven by government assistance (e.g. R&D subsidies and grants); collaboration with industry association and government agencies that is driven by new product development and government assistance; collaboration with universities which is driven by new product development, R&D investments and government assistance; collaboration with external/private laboratories and hospitals which is driven by new product and new process development, and government assistance; and collaboration with traditional medical practitioners which is driven by new product and new process development.

Public R&D organizations have traditionally been funded by the state, and these organizations had in the past no incentive to commercialize their R&D results. However, the initiative to transform them into financially autonomous entities and for them to adapt to the new model of operation will take some time. Moreover, most R&D organizations did not have technology transfer offices or a special office to carry out these kinds of activities. Many organizations had in the past – for free – transferred to farmers or other users their R&D results but they did not have mechanisms of transferring to industry for profit.

As with public R&D organizations, Vietnamese enterprises also display considerable socially negative habits developed over the long period of a centrally planned economy, which bred many state-owned enterprises. Although many state-owned enterprises are currently being equitized, they still lack experience of a market-based economic system. As to be expected, many firms are not active in innovation and do not have any connection with R&D sector. For this reason there is a lack of a mechanism to exchange information between the R&D sector and firms, and between enterprises and government agencies. Many firms even do not have information on incentive policies offered by the state, and for much of the time lack information and knowledge about the laws and regulations of overseas markets that are very important for the export market, especially for those in the agriculture sector (Nguyen Vo Hung 2003, Nguyen Thanh Tung, 2004).

Another set of institutional weakness is that Vietnamese enterprises operate without appropriate standards; for example many firms in the field of food processing in particular and in agriculture in general, are unable to meet the standard requirements of high demanding markets such as EU and Japan. Concerning training, most universities in Vietnam focus on training, due to lack of equipment for experimental work, while the students tend to master theory, not practices.

Intellectual property rights (IPRs) are not perceived as important not only in industry but also by public R&D organizations which lack knowledge of IPRs and related issues. The number of inventions patented in Vietnam is very small in comparison with many countries in the region and in the world (about 2000 applications a year). Moreover, most patents are granted to foreigners; the number of patents granted to Vietnamese account for about 10%. R&D organizations hardly apply for intellectual property protection until recently, when enterprises started to pay attention to IPRs issues but do so more about trademarks.

4.8   State policies and regulatory framework for biotechnology development

Key incentives and policies for enhancing Vietnamese biotechnology development set out in Resolution 18/CP are described below.

Enterprises are given incentives to engage in innovation by implementing R&D activities or importing advanced technology for processing new biological products, which significantly promote socio-economic development. The government provides incentives such as tax reduction, a preferred interest rate, favourable procedures, and other institutional support to these enterprises, to products for which the technology are locally available. Energy-intensive products and those potentially harmful to human health and environment are being explicitly discouraged.

A National Committee on Biotechnology was established in 1997 to advise the Prime Minister in the implementation of Resolution 18/CP on biotechnology development. In order to implement the resolution, as indicated earlier, the Ministry of Science and Technology (MOST) established state-level research programme, titled 'Biotechnology for developing sustainable agriculture, forest and aquaculture, environment and human health protection' (1996–2000) and have been implementing the research programme, titled 'Science and technology for biotechnology development' (2001–5). Simultaneously, MOST has built facilities for 60 laboratories related to biotechnology based at institutes, universities, and six national biotechnology laboratories all over the country.

In terms of human resource training, the Ministry of Education and Training (MOET) have promoted the training of biological human resource. Largely through cooperation with the MOST, which implement various research programmes. This, collaborative training emphasizes the two pillars of human capabilities and innovation infrastructure both of which are important factors for developing Vietnamese biological system. In additional, MOET is running a programme that provides funding for selected students to study abroad; by investing about US$6 million every year, by 2003, 30 students and 200 staff have been funded to study overseas specializing in biotechnology (Quan, 2004).

In 2000, the MOST and other ministries agreed on a strategy for developing biotechnology as part of the national S&T strategy which was approved in 2003. Apart from the overall policy document, a range of other more specific government policies on crop and animal varieties, incubating vital young plants, state-supported agricultural contracts for high-tech agriculture, forest and aquaculture development programmes have been promulgated. Despite all these efforts, current policies are deemed by most actors to be unclear and not specific enough to make an impact on biotech development.

To further strengthen the system, the Prime Minister approved a programme on biotechnology development and application in the area of agriculture and rural development. This programme – proposed to run until 2020 – aims to create new plant varieties, animal breeds, micro-organic strains and agricultural biotechnological preparations of high yields, high quality and economic efficiency in order to well serve economic restructuring in agriculture and rural development; to raise the quality and competitiveness of commodity farm produce and the proportion of processed agricultural, forestry and aquatic products in service of domestic consumption and export. An amount of US$62.9 million was budgeted to develop and apply biotechnology in agriculture and rural development over the 15-year period. Funding will be provided for scientific research, trial productions, and specialized training for the programme, while the Ministry of Agriculture and Rural Development will to monitor this programme.

To promote international cooperation in biotech, the government has encouraged organizations to use international experts or Vietnamese citizens living abroad with high professional qualifications in order to act as consultants or to take part directly in the development plans, training processes, and R&D activities. In some specific cases, the government has encouraged and financed bilateral and multi-lateral cooperation between Vietnam and international organizations in biotechnology. Apart from the legal documents mentioned above, government policies on crop and animal varieties, incubating vital young plants, state-supported agricultural contracts for high-tech agriculture, forest, and aquaculture development programmes have been promulgated. The policies for encouraging investment in all economic sub-sectors that promote high-tech agriculture will continue to be supported using institutional mechanisms such as infrastructure investment, tax reduction, import tax exemption on imported equipment, among others.

In general, in order to implement the broader policy indicated under Resolution 18/CP to promote Vietnamese biotechnology development a number of institutional and regulatory steps have been taken. The resolution, according to different actors, has had good impact on the sector, although, the results are still limited due to limited finance invested in this field by the state as well as by other actors. Beside the biotechnology sector has not developed as expected because of the weaknesses of the innovation system: linkages between actors of the system are very weak, while there is limited competence of actors. Most actors of the sector are still affected by habits and practices developed from the centrally planned economy; however, it is too early to assess the behavioural impact of actors on the many policies, laws, and regulations that are being enacted. In spite of the myriad legislations, Vietnam still lacks laws and regulations regarding to biosafety, which is important for the development and operation of biotechnology sector.[7]

4.9   Summing up

Over a relatively short time, driven in large part by government efforts a broad range of infrastructure, organizational, and institutional frameworks to develop capabilities in biotechnology have been created in Vietnam which has led to improvement in research and innovation capacity. Applications in production have contributed to improvement of the quality of agricultural and aqua cultural products; there have been improvements to the income of farmers as well as creation of jobs in several agricultural sub-sectors. However, Vietnam is taking only the first steps toward the development of modern bioscience and biotechnology. According to an assessment by the Biotechnology Atlas Project (MOST, 2003), among the ten ASEAN countries, Vietnam is ranked in the bottom half. Biotechnology in Vietnam lags behind other regional competitors and currently lacks the innovation capacity to meet the increasing demand of a myriad of socio-economic challenges and to improve living standards.

In specific terms we examined systematically the role of state institutions in producing new product and/or process development, the dynamics of R&D investments over the period 2001–5 and the determinants of collaboration in the Vietnamese biotechnology sector. The results are summarized as follows: first, IPP, the quality of local infrastructure services and the participation in local SMI development schemes affect positively and significantly new product and/or process development while the availability of venture capital and the transfer of personnel to local firms or R&D institutions (for training) play a negative role in new product and/or process development. Second, there is evidence of 'true' persistence in R&D investments in the sense that, after controlling for the individual effects and solving the initial conditions problem, past R&D investments strongly, positively and significantly affect current R&D investments. Current employment also affects positively and significantly current R&D investments and unobserved heterogeneity plays a highly significant role in current R&D investments. Finally, we found that, in general and regardless of the type of collaboration, collaborators and non-collaborators have similar characteristics. However, collaboration with PRIs is driven by government assistance (e.g. R&D subsidies and grants); collaboration with industry association and government agencies is driven by new product development and government assistance; collaboration with universities is driven by new product development, R&D investments and government assistance; collaboration with external/private laboratories and hospitals is driven by new product and new process development, and government assistance, and collaboration with traditional medical practitioners is driven by new product and new process development.

In sum, a number of issues that are relevant to innovation capacity in biotech development were thrown up. First, local and international cooperation is a crucial starting point for research and innovation and this thrives only within a conducive environment to enable manufacturers use knowledge from the research sector. No less important is a market incentive regime that stimulates entrepreneurship of scientists. In this regard Vietnam is formulating policies to attract multinationals, first to build industrial production capacity and, in time, hopefully related R&D capacity. The law on technology transfer currently being drafted is expected to contribute to a better environment. Efforts to restructure the whole law and legal regulations as well policies for joining WTO is another important factor that would affect the sector and as well, the recently enacted IPR Law and other regulations on biosafety should create a new environment for a more dynamic innovation system.

Again, although the Strategy for Vietnamese S&T development has been approved, where the priorities in biotechnology and bio-industry were defined, these strategic objectives are not fully specified; this is necessary to further narrow priority fields in R&D and biotechnology applications. Biotechnology is an expensive activity for latecomer economies, which need to focus on a few activities and find the niche where the country's bioscience and biotech can best address local needs, as the experiences of Cuba, South Africa, and elsewhere show the importance of addressing local needs.

Third, learning is a key ingredient for innovation success; this includes learning from international partners, as well as domestic internal learning from local actors. One of the measures to address the fragmentation of investment, build up of biotech innovation system, as well as breakdown institutional barriers is the creation of centres of excellence in bioscience and biotech. The Millennium Science Initiative (MSI) with the involvement of the World Bank is being examined as an opportunity to address this issue in Vietnam.

The analysis of biotech innovation system in this chapter reveals that actors from government, international cooperation mechanisms and market driven forces are all important. The role of private actors and entrepreneurs with the right management skills in the biotech activities will become increasingly important.

Appendix

Vietnam: The persistence of R&D investments
We consider a dynamic panel data type 1 tobit (according to Memiya's (1984) terminology) that explains current R&D investments as a percentage of total sales by past R&D investments and current total employment. The model allows us to control for unobserved heterogeneity through the individual effects.

Formally, it consists of a latent variable $\overset{*}{y}_{it}$ written as

$$\overset{*}{y}_{it} = \gamma y_{i,t-1} + \beta' x_{it} + \alpha_i + \varepsilon_{it},\tag{1}$$

with $i = 1,...N$; $t = 1,...T$ ($N = 66$ and $T = 5$, from 2001 to 2005). The latent variable capture the incentive to invest in R&D which is a function of past R&D investments $y_{i,t-1}$, current total employment x_{it},[8] individual effects α_i and other time-varying unobserved variables ε_{it} . If the incentive to invest is high enough, i.e. $\overset{*}{y}_{it} > 0x_3$, firm I does invest in R&D at period t, i.e. current R&D investments are observed to be positive, otherwise they are zero. Formally, the observed counterpart to $\overset{*}{y}_{it}$ is written as

$$y_{it} = 1[\overset{*}{y}_{it} > 0]\overset{*}{y}_{it},\tag{2}$$

where $1[...]$ is the indicator function with value one if the expression between square brackets is true, and zero otherwise.

We assume random effects distributed according to the normal distribution and correlated with the explanatory variable. ε_{it} is also assumed to be normal so that the model is estimated by maximum likelihood. The likelihood function is given in Hsiao (2003) and involves a one-dimensional indefinite integral which is evaluated using adaptive Gauss-Hermite quadrature along the lines of Rabe-Hesketh et al. (2005). The so-called initial conditions problem, due to the dynamic feature of the model, is solved using Wooldridge's (2005) 'simple solutions' that consist in writing the individual effects, in each period, as a linear function of the explanatory variable and the initial conditions, i.e.,

$$\alpha_i = b_0 + b_1 y_{i0} + b_2' x_i + a_i,\tag{3}$$

where $x_i' = (x_{i1}', ..., x_{iT}'), b_0, b_1$ and b_2' are to be estimated, and a_i is independent of (y_{i0}, x_i).[9] The scalars b_1 captures the dependence of the individual effects on the initial conditions.

Table 4.6 Descriptive statistics of employment, domestic inputs and R&D over the period 2001–5

Variable	Mean	(Std. err.)	Min.	Max.
2001				
Total employment (FTEs)	286.682	(973.095)	1	7500
% of gross inputs sourced domestically	31.152	(40.097)	0	100
R&D expenditures (% of total sales)	2.783	(9.198)	0	65
R&D staff (% of total workforce in FTEs)	3.600	(11.765)	0	73
2002				
Total employment (FTEs)	297.848	(1032.119)	1	8000
% of gross inputs sourced domestically	33.977	(40.878)	0	100
R&D expenditures (% of total sales)	3.942	(11.084)	0	65
R&D staff (% of total workforce in FTEs)	3.381	(9.734)	0	50
2003				
Total employment (FTEs)	334.182	(1016.457)	1	8000
% of gross inputs sourced domestically	40.152	(41.604)	0	100
R&D expenditures (% of total sales)	4.509	(10.95)	0	65
R&D staff (% of total workforce in FTEs)	4.117	(12.087)	0	73
2004				
Total employment (FTEs)	451.864	(1004.819)	1	8000
% of gross inputs sourced domestically	50.758	(40.307)	0	100
R&D expenditures (% of total sales)	5.619	(11.188)	0	65
R&D staff (% of total workforce in FTEs)	4.626	(10.92)	0	53
2005				
Total employment (FTEs)	505.485	(1119.47)	1	9000
% of gross inputs sourced domestically	58.883	(36.466)	0	100
R&D expenditures (% of total sales)	5.765	(10.892)	0	70
R&D staff (% of total workforce in FTEs)	7.148	(13.119)	0	73
Number of firms		66		

Table 4.7   Correlation between the types of collaboration

	Public research institutions	Industry assoc.	Universities	External/private laboratories	Hospitals	Traditional med. practitioners	Plant org.	Farmer groups	NGOs	GMOs org.	Envir. org.	Govt. agencies
Public reasearch inst												
Public reasearch inst	1.000											
Industry assoc.	0.543**	1.000										
Universities	0.637**	0.514**	1.000									
External/private lab.	0.412**	0.525**	0.373**	1.000								
Hospitals	0.264*	0.301*	0.223†	0.323**	1.000							
Trad. med. pract.	0.097	0.287*	0.111	0.280**	0.585**	1.000						
Plant organizations	0.136	−0.029	0.019	0.060	−0.262*	−0.311*	1.000					
Farmer groups	0.114	0.083	0.272*	0.024	−0.239†	−0.300*	0.673**	1.000				
NGOs	0.265*	0.077	0.207†	−0.019	−0.022	0.008	0.319**	0.499**	1.000			
GMOs organizations	0.007	0.020	0.014	0.093	−0.027	−0.103	0.331**	0.343**	0.216†	1.000		
Envir. organizations	0.067	0.031	0.085	0.158	−0.262*	−0.188	0.290*	0.391**	0.394**	0.371**	1.000	
Govt. agencies	0.338**	0.363**	0.315**	0.422**	0.285**	0.201	0.256*	0.239†	0.108	−0.030	0.205†	1.000

Significance levels: † : 10% * : 5% ** : 1%

5
Nigeria as a Very Late Follower in Agricultural Biotechnology

5.1 Introduction

Agriculture has been a major contributor to the Nigeria's economy, accounting for about 40% of gross domestic product (GDP) and employing about 60% of the work force. However, since 1973 the economy has been highly dependent on the oil sector, which by 2006 accounted for 70–80% of government revenues, over 90% of export earnings, and 25% of GDP, (Central Bank of Nigeria (CBN), 2007). Local agriculture employs more than 90% of the rural population with women playing a major role in the production, processing, and marketing of food crops. Arable land is estimated at more than 70 million ha, of which only 50% is under cultivation. Agricultural land covers a wide range of agro-ecological zones ranging from the semi-arid regions of the northern-most states, the derived savannah land of the middle belt and south-west, to the rainforest belt of south-south and south-east. The total area of inland water bodies is estimated at slightly above 12 million ha. Low-lying and seasonally flooded areas increasingly produce cereals such as rice. Forests and woodlands occupy 17 million ha, but primary forests and most of the wildlife are disappearing.

Like much of sub-Saharan Africa (SSA), Nigerian agriculture is predominantly smallholder-based with over 95% of the output from smallholdings ranging in size from 1 to 5 ha. According to Nigeria's National Bureau of Statistics (NBS) data, about 45–57% of farmers grow only food crops while the rest grow food and cash crops (NBS, 2007). The agricultural produce comprises several staple crops arising from the diversity of agro-ecological production systems. The major food crops are: cereals (sorghum, maize, millet, and rice); roots and tubers (yam, cassava, and potato); legumes (groundnut and cowpeas), and others (fruits and vegetables). These commodities are of immense importance for food security, household income, employment, expenditure, and scientific and technological research.

Among the above staple crops, cassava has assumed a prime position for the country, with Nigeria now becoming the world's largest producer ahead of Brazil and Thailand. In spite of its relative success in feeding its population, Nigeria faces the broad constraints of a very latecomer economy that relies on rain-fed production subject to high seasonal fluctuation, uneven product quality and variations in supply and prices, low productivity, and high production costs. Nigeria suffers from low agricultural productivity, which can be attributed to the lack of scientific solutions to a host of tropical agro-ecological characteristics, the most significant being poor soil structure in some areas of the country, pests, diseases, and drought. The structural challenges include low-level, rural telecommunication, inadequate and poor feeder roads, a lack of good quality research tailor-made to address local agricultural needs, high transportation costs, and the absence of good agricultural extension services. Additionally, the poor market information system and poor linkage between producers, processors, traders, and consumers all point to a characterization of the country's agricultural system as a very latecomer in our characterization of country capabilities according to Chapter 2.

While this situation has compelled the Nigerian government to take interest in agricultural biotechnology as a means to improve local food production, there is as yet no biotechnology crop under development in Nigeria despite several years of policy discussions, debate, and public-sector activities. This chapter presents a critical analysis of the state of the sectoral system of agricultural biotechnology in the country, bearing in mind precisely this disjuncture between policy aims and ground realities. The data presented in this chapter was collected during a 2006 survey of universities and PRIs in Nigeria on aspects of collaborative behaviour within the system of innovation. Although there have been surveys on the activities of researchers in biotechnology research and innovation in Nigeria (Ilori et al., 1994; Irefin et al., 2005), this is the first instance of a comprehensive survey of public-sector activities related to agricultural biotechnology in the country. A total of 210 questionnaires were retrieved out of the 250 administered and face-to-face interviews were carried out with more than 50 scientists and high-ranking government officials. Within universities, our units of measurement included mainly the departments of agriculture, chemistry, biology, biotechnology, biochemistry, and microbiology. Firms operating in the enterprise sector were not surveyed because local private-sector capacity to exploit biotechnology-based innovations for agriculture is almost non-existent in Nigeria.[1]

Section 5.2 discusses the key actors in the agricultural biotechnology system involved in knowledge generation activities. This is followed by an analysis of the kinds of science, technology, and innovation (STI) investment that are ongoing in the country, and the major innovation constraints, namely, human resources, funding, and the lack of ancillary organizations. Section 5.4 presents our data on the factors that promote interactive learning and collaboration within the sectoral system, paying particular attention to the kinds of collaboration, the incentive structures, and the relative payoffs that are observable in terms of actor behaviour and networks. How this is linked to the policy regimes governing agricultural biotechnology and the role of the state in championing agricultural biotechnology's potential for local solutions is dealt within section 5.5. We follow this with a case study of cassava (see section 5.6) that has been so successful in order to see if there are any divergences from the general trends in the case that caused its huge success and, if so, what lessons there may be to share.

5.2 Key actors generating agriculture biotechnology innovations and knowledge

Nigeria has set up several knowledge-related institutions (such as PRIs and centres of excellence) since the colonial period, however, there has been poignant deterioration due in part to a lack of investment in facilities and human capacity development under its military regime (IFPRI, 2006). By the end of the 1990s, the public-sector infrastructure had evidenced a massive decline both within universities and PRIs that can be ascribed to a multitude of factors, namely:

(a) An unstable political situation and declining support from governmental agencies.
(b) The Structural Adjustment Programme (SAP): reduced educational expenditure, one of the pronounced negative effects of the SAP.[2]
(c) A lack of cooperation with other actors within and outside African countries.
(d) Expanding university enrolment and graduate output: universities within Nigeria have had expanding enrolments with no or little by way of a commensurate increase in academic staff, research funding, and other infrastructure.[3] The expenditure per student at the

university level declined drastically between 1984–85 and in 2000 throughout the country (Oyeyinka et al., 2005).

5.2.1   Nature of knowledge activities

Within the range of technological applications forming part of agricultural biotechnology, our survey showed that most of the researchers in both universities and PRIs were engaged in low knowledge-intensive aspects of biotechnology. In fact, cell and tissue culture appear to be relatively more common than any other types of biotechnology research across all organizations surveyed. While more than four-fifths (84%) of the researchers that took part in the survey were engaged in cell and TC, only two-fifths of the 210 respondents worked with relatively more advanced biotechnologies such as recombinant DNA or genetic engineering; and less than one-third were involved in bioprocess technologies and bioinformatics; while only 15.8% of the researchers worked in molecular diagnostics. The data shows that cell and tissue culture is the most important focus of biotechnology research according to 52.6% of the respondents; while 26.3%, 10.5%, 5.3% and 5.3% claimed bioprocess technologies, molecular diagnostics, bioinformatics, and genetic engineering, respectively, as the most important focus of their biotechnology research.

Table 5.1 below shows biotechnological tools and areas of work of two major PRIs (NIPRID and SHESTCO Complex) within the country from an earlier survey conducted by us in 2003–4. Our results are confirmed by other authors who have worked on understanding how resources are allocated to various knowledge-based activities within agricultural biotechnology in the country (see Mugabe, 2000; Adeoti, 2002).

The nature of knowledge activities is linked to the availability of human skills locally, a point that is discussed in detail in the next section. Nevertheless, the range of activities   prevalent locally in agricultural biotechnology make a preliminary case that Nigeria is quite low on the overall knowledge intensity of activities in this sector.

5.2.2   Main actors

Universities and PRIs are the main drivers of biotechnology-based innovation in Nigeria as evidenced by the survey. Although we found several subsidiaries of international companies, we were not able to locate any that conduct significant R&D activities related to agricultural biotechnology within the country. Several of them have microbiology laboratories on their premises, but these were largely

Table 5.1 Biotechnology tools and areas of work: Results of the 2003 survey

Institutions/ activities	Percent of financial resources devoted	Average number of researchers	How long work has gone in this area?	% of your total activity in 2003
NIPRID*				
Cell and TC	17.5	3.0	3.0	11.3
Recombinant DNA	20.0	3.7	5.5	33.3
Molecular diagnostics	61.3	12.3	8.3	
Bioinformatics				
Bioprocessing				57.2
SHESTCO**				
Cell and tissue culture		2.0		33.3
Recombinant DNA				6.7
Molecular diagnostics		2.0	8.5	
Bioinformatics				
Bioprocessing				60.0

Source: Field survey by authors (2003–4).
Note: * National Institute for Pharmaceutical Research and Development.
** Sheda Science and Technology Complex, Abuja

employed for minor testing and other activities.[4] The local private sector is conspicuous by its absence, mainly due to low levels of entrepreneurship and poor institutional capacity support for translating inventive effort into innovation. Our survey found the existence of distribution-based firms but these are not engaged in any knowledge-related activities.

The international Institute for Tropical Agriculture (IITA) based in Ibadan is part of the Consultative Group on International Agricultural Research (CGIAR) system (and hence not a local PRI) but counts as perhaps the foremost organization where sustained research using modern agricultural biotechnology methods is being conducted in the country. The institute was doing preliminary work on the genetic transformation of cowpea and cassava at the time the survey was being carried out. IITA also collaborates with several national institutes in this area and helps build capacity by training skilled manpower in laboratories equipped to perform advanced biotechnology-based research. For example, IITA has collaboration on-going with the National Root Crops Research Institute (NRCRI) on biotech cassava research as well as with other national actors; exemplified by the fruitful collaboration to develop the cassava production in Nigeria discussed in section 5.6.

5.3   Science, technology, and innovation (STI) investments

The emergence of agricultural biotechnology in Nigeria is perhaps most closely associated with research and policy triggers at the national level and the availability of sporadic international research funding. Initiation of certain national policies and programmes related to agricultural biotechnology, no doubt lent strength to initiate activities, but STI investments continue to be very weak and do not provide a basis for a wider knowledge base required to conduct advanced innovative activity in agricultural biotechnology of relevance to local food security. The most important innovation constraints listed by survey respondents are: human resources, sustained funding for scientific infrastructure, and a lack of ancillary institutions.

5.3.1   Human resources

The predominance of cell and tissue culture activities among organizations active in agricultural biotechnology is a cumulative result of a lack of trained personnel to conduct more knowledge-intensive activities, on the one hand, and funding issues that prevents them from venturing into higher domains, on the other. The lack of skills for biotechnology-based work finds evidence in the literature as far back as 1994, where Okafor reports that Africa had only 106 trained modern biotechnologists; of these Nigeria had about ten scientists trained in gene cloning (Okafor, 1994). In a survey of 17 universities and agricultural research institutes in Nigeria, Alhassan (2001) found that Nigeria had only 65 highly trained personnel in core molecular biology, 58% of whom are Nigerian PhD holders. In addition to these, 34 PhD holders are available from ancillary biological fields with supplementary training in molecular biology. A disproportionately large number of these (34) were from the field of microbiology. In a similar survey (Irefin et al., 2005), out of a total of 35 biotechnologists, 18 scientists had Masters degrees while 14 had doctoral degrees.

Our survey found that while most scientists specialized in conventional biotechnology and plant tissue culture (TC) techniques, finding skilled personnel who were well-versed with r-DNA technologies, cell-fusion technologies and other advanced genomics-related training was far more challenging. The production of planting material (through constant replication techniques) is the most active area of application of biotechnology followed distantly by diagnostics, bio-fertilizer, and vaccine production.[5]

Combined with earlier surveys aimed at assessing biotechnology capacity of Nigeria for agricultural research in universities, polytechnics, and national research institutions (Duduyemi and Omitogun, 2004), we provide a somewhat tentative list of organizations undertaking biotechnology research in Nigeria. Although there has been a historical deterioration of scientific facilities in most universities as we noted in our survey, there is the potential for a number of institutions capable of applying some biotechnology tools to improve agricultural production in Nigeria. Nigeria has the base for producing the requisite research scientists but acquisition of specific skills as well as the facilities and tools for biotechnology requires that the state make the financial investment.

Below we show the current focus of each institution's biotechnology research activities. Box 5.1 contains a list of the most prominent organizations involved in biotechnology in the country.

Box 5.1   Organizations engaged in biotechnology research In Nigeria

1. *University of Agriculture Abeokuta (UNAAB) Biotech Center:* Training, marker-assisted breeding, biodiversity
2. *Obafemi Awolowo University (OAU)*: Animal Biotechnology Laboratory, biodiversity, characterization, conservation
3. *National Center for Genetic Resources and Biotechnology (NACGRAB)*: Biodiversity, seed conservation, gene bank storage
4. *Federal University of Technology, Akure (FUTA)*: Feed development and enhancement, diagnostics
5. *Ladoke Akintola University of Technology (LAUTECH)*: bio-fertilizers, feed development
6. *University of Lagos (UNILAG)*: Bioremediation
7. *NAPRI:* breeding
8. *National Veterinary Research Institute (NVRI), Vom*: Vaccine production
9. *Federal Polytechnic, Ado-Ekiti*: Feed and food processing, alternative feeds (plasma protein, blood meal)
10. *University of Ilorin (UNILORIN)*: Diagnostics, marker-assisted breeding
11. *Federal Environmental Protection Agency (FEPA)*: Biogas, insecticide and acaricide extractions

12. *University of Ibadan (UI) Department of Veterinary Medicine:* Diagnostics
13. *University of Benin (UNIBEN):* Biodiversity, marker-assisted breeding
14. *Federal Institute of Industrial Research Oshodi (FIIRO):* Aflatoxin, food and feed fermentation
15. *Shedu Science and Technology Complex, Abuja (SHEDA), Abuja:* TC
17. *National Biotechnology Development Agency (NABDA):* research into reproductive rate of grass cutter (thryonomys spp, utilization of toxins from poisonous mushroom, disease resistant research of potatoes and papaya; mandate: genetically modified soybeans, corn, cotton, sorghum, banana/plantain, and so on).

5.3.2 Funding

We found that funding impacts both physical and scientific infrastructure in extremely critical ways. Most of the biotechnology laboratories visited during the survey did not have access to a stable electricity supply and clean water. The already constrained resources for research efforts were further split between ensuring the availability of such basic infrastructure and providing research inputs.[6]

Survey respondents affirmed the limitations of funding and were most concerned by the impact of a lack of sustained funding commitments by the government on conducting R&D in agricultural biotechnology. Most of them complained about having to rely on external (read donor-based) funding in order to be able to perform any research of significance at all within their premises, and how this often compelled them to shift gears completely from their basic mandates, since donor priorities do not often match with those of the country's own science and technology vision. Limited funding has resulted in large-scale variations in biotechnology-related work among the universities and PRIs. All the organizations surveyed showed significant differences in their work on cell and tissue culture and advanced activities in biotechnology. Whereas the differences in advanced activities in the sector is because of lack of good knowledge base and collaborative venues (see next section), the former suggests that there is a large variability in the quality of laboratory facilities across the public sector even to conduct some basic biotechnology research.

The survey also found that several international funding initiatives in agriculture biotechnology were aimed at strengthening local capacity for biosafety and enhancing public knowledge of genetically modified (GM) crops and not directly targeted at expanding the innovation base of the local organizations. For example, agricultural biotechnology in Nigeria received a boost with the launch of two linked initiatives funded by the United States Agency for International Development (USAID) in 2004: the West African Biotechnology Network (WABNET) and the Nigeria Agricultural Biotechnology Project (NABP), implemented by the IITA in close collaboration with Tuskegee University. The National Agricultural Biotechnology Project was a US$2.1 million project to assist Nigeria in building the framework for decision-making that will facilitate access to the opportunities biotechnology offers and will ensure the safe and effective application of this technology to improve agriculture. A key element of the project is to improve implementation of bio-safety regulations; and, enhance public knowledge and acceptance of biotechnology. Similarly, Eicher (2005) reports biosafety application that was concluded by the Danforth Centre in collaboration with IITA, the National Biotechnology Development Agency (NABDA), and the National Root Crops Research Institute (NRCI).

Table 5.1 presents some descriptive statistics from the field survey on the issues under consideration. Of the 210 respondents, only 27% thought that their activities qualified to be called product innovation, whereas 26% felt that they were involved in process innovation. A total of 30% of the survey respondents had overseas training in research of relevance to agricultural biotechnology; while about twice as many have only received local training. The impact of physical infrastructure on innovation activities was also captured. Most organizations felt that the presence of power supply and other infrastructure hardly contributed in any positive way to their innovation efforts, since they could never rely on it (table reports that the number of respondents who felt that the present state of infrastructure contributed positively to their innovation endeavours in fairly strong, strong, or very strong ways was very small (9% and 3% respectively). Only 40% of the organizations have research funding sources from the government, and only 19% of them were of the opinion that government incentives played a role in product and process innovations. Both types of innovation, namely

Table 5.2 Descriptive statistics: Innovation, learning, and government policies and funding

Variable	Mean	(Std. dev.)	Min.	Max.
Product innovation	0.271	(0.446)	0	1
Process innovation	0.262	(0.441)	0	1
Human capital	0.514	(0.501)	0	1
Foreign training programmes	0.314	(0.465)	0	
Local training programmes	0.748	(0.435)	0	1
Government innovation incentives	0.186	(0.390)	0	1
Skilled manpower	0.567	(0.497)	0	1
Technical collaboration	0.581	(0.495)	0	1
Laboratory facilities	0.229	(0.421)	0	1
IP protection	0.267	(0.443)	0	1
Quality of ICT	0.338	(0.474)	0	1
State of power supply	0.090	(0.288)	0	1
State of water supply	0.190	(0.394)	0	1
Other policies	0.029	(0.167)	0	1
Government funding	0.396	(0.417)	0	1
N		210		

Source: Field survey by authors (2006).

product and process innovation are correlated, albeit to a small extent (0.221), significantly at 1% level of significance.

5.3.3 Ancillary organizations

Given the absence of the private sector to a large extent, the survey sought to ascertain the establishment of ancillary organizations that could perform some or the other functions such as state-owned enterprises, to boost uptake of research results and apply it to production; venture capital or other forms of funding programmes and agencies; and technology transfer offices that could liaise with the universities and PRIs to help them articulate their knowledge-based requirements. The survey found no systematic efforts by the government to invest in any of these ancillary STI mechanisms that whose use and applicability for agricultural biotechnology is widely documented in the innovation studies literature.

5.4 Interactive learning and collaboration

The low base of knowledge activities in the sector, as determined by: (a) the availability and configuration of human skills, (b) funding of scientific and physical infrastructure, and (c) policies and institutions that condition actor interactions – all impact on how actors interact and what forms of learning take place in the course of research and knowledge production activities.

In this section, we use survey data to analyse the patterns of collaboration evident in the local sector, the government policies (or lack thereof) that trigger collaboration (or obstruct it), and the impact of networking on sectoral performance.

5.4.1 Types of collaboration

Table 5.2 presents descriptive statistics for technological performance and the networking explanatory variables. Once again, the survey data shows that the percentage of organizations that have a commercial success regarding their product innovations is rather small (10.5%) and most of them deem the collaboration intensity with universities to be fairly strong, strong, or very strong (53.3%), compared to collaboration with other types of institution (e.g. 11.9% with agricultural machinery suppliers). This is an interesting finding in its own right because it corroborates a very important issue in innovation systems in late development: that of informal attitudes of researchers/university scientists. Probing further into nuances of actor behaviour, survey respondents, especially university academics do not conceive their role in entrepreneurial terms while research institutes tend to be better disposed towards getting their research commercialized. However, without exception, their efforts are undermined by a host of factors which include not just poor infrastructure but in a more fundamental way socially rooted attitudes and norms that are reinforced by the institutional framework under which the organizations were set up in the first place. On the question of poor infrastructure, university academics mostly admitted to conducting their research work during academic sabbaticals abroad. International research grants seem to be playing a predominant role in enabling the staff to maintain existing laboratory facilities, but in the absence of proper basic infrastructure (electricity, etc), laboratory facilities and money for consumables, and the researchers find it very difficult to take research forward. However, more important than this seems to be the unwritten yet widely accepted, deep-rooted norm that universities do teaching and research and the private-sector firms do product development. The notion that university research should feed systematically into national production has never been promoted and in the extreme, academics have a propensity to view the idea as hostile. This is the reason why although the survey captures a high rate of collaboration between universities and PRIs, these do not result in greater innovation performance of the sector as a whole.

Both universities and research institutes collaborations can be classified broadly into teaching, research, and product development categories. Descriptive statistics of these kinds of collaboration with both local and foreign partners is reported in Table 5.3 below. The figure shows the mean results in an interval between 0 (weak) and 1 (strong) where collaborations are measured in binary variables indicating whether a firm has foreign or local collaboration to effect production, in research, teaching, product development, in R&D, and in extension services. Hence, any result above 0.5 shows moderate to strong correlation.

Closer scrutiny of the collaborations reveals that more organizations are involved in local collaborations than in foreign collaborations (the mean for foreign collaboration is relatively low compared with local collaboration in research – 0.086 versus 0.605 as reported in Table 5.3). But despite this, the organizations involved in foreign collaborations are more likely to commercialize product innovations than those involved in local collaborations, as Table 5.4 shows. This could be attributed to two factors. First, most foreign collaborations that were ongoing at the time of the survey were aimed at small and concrete innovations, often even training components, and did not amount to large-scale product development of new seed varieties. Second, the foreign collaborations tended to be better structured, evenly funded through all phases of the project in question, and there were clear demarcation of roles of responsibilities of all partners. In contrast, the local collaborations tended more to be affected by the distortions in the local innovation systems, especially those related to funding and

Table 5.3 Descriptive statistics: Technological performance and networking

Variable	Mean	(Std. dev.)	Min.	Max.
Technological performance	0.105	(0.307)	0	1
Col. with research institutes	0.362	(0.482)	0	1
Col. with farmers associations	0.376	(0.486)	0	1
Col. with universities	0.533	(0.500)	0	1
Col. with private laboratories	0.176	(0.382)	0	1
Col. with external agencies	0.295	(0.457)	0	1
Col. with agricultural machinery suppliers	0.119	(0.325)	0	1
Col. with agricultural cooperatives	0.152	(0.360)	0	1
Col. with seed companies	0.157	(0.365)	0	1
Col. with others	0.043	(0.203)	0	1
N		210		

Source: Empirical survey by authors (2006).

Table 5.4 Foreign and local collaborations

Variable	Mean	(Std. dev.)	Min.	Max.
Foreign collaborations in production	0.086	(0.281)	0	1
Foreign collaborations in research	0.319	(0.467)	0	1
Local collaborations in research	0.605	(0.49)	0	1
Local collaborations in teaching	0.176	(0.382)	0	1
Local collaborations in product development	0.414	(0.494)	0	1
Local collaborations in R&D	0.414	(0.494)	0	1
Local collaborations for extension services	0105	(0.105)	0	1
N		210		

Source: Empirical survey by authors (2006).

insufficient incentives in the policy framework to structure mutually beneficial collaborations. The survey also found that organisations had obsolete mandates, several of which overlapped with one another, and tended to compete with each other for importance and funding within the innovation system, which contributed to a lack of collaborative spirit. As Table 5.3 shows, local collaboration for both product development and R&D are just slightly below average (0.4) but these tend to be not directed towards creating marketable products.

A probit estimation conducted on the basis of the survey data on other variables that are statistically significant for product development in addition to foreign collaboration (not reported here) shows that: (a) collaborations for product development and for local research, and (b) presence of extension services are both positively and significantly correlated to innovation performance. The probit analysis also shows that local collaboration in R&D is, however, significant but negative. This helps to capture the poor impact of R&D on collaboration and innovative performance in agricultural biotechnology in Nigeria and once again, strengthens the point that collaboration with one/two of system actors is insufficient for the sectoral system to learn and bear the fruits of innovation, a point made in the earlier section.

5.4.2 Determinants of collaboration

Our interviews show that collaboration has been limited by three main factors, namely:

- The inability of scientists to move their work beyond the individual organizations;
- The absence of formal institutions supporting collaboration, and;
- Poor incentive to motivate scientists.

The institutions in which scientists work barely reward entrepreneurship and there is no motivation to make additional effort beyond the publication of academic papers. Most academics do not understand the institution of patents for instance and have had little guide as to what to do to move inventions to the market. Moreover, the sheer weight of infrastructural constraints leaves them with very little energy to think beyond their immediate concerns. The concerns with short-term goals was evident and long-term commercialization efforts was way down the priority list of scientists. It was evident to the research groups that major institutional shifts would be required to change the present habits and practices. While individual researchers are able to carry on working up to a point, the odds rise dramatically as projects demand better facilities, skills, and knowledge that the lone scientist could offer.

We also found formal institutional arrangements that could support a positive research culture and also focus attention of researchers on applied research that leads to product development were missing in almost all cases. This includes the presence of human skills, ancillary institutions, and appropriate incentives for collaboration. We also concluded that collaborative efforts between the university departments/ PRIs and external technical partners that do not necessarily focus on immediate local needs are inevitable, since they are one of the few venues of intellectual engagement for scientists and researchers.

We identified three factors why the innovation process remains at the level of pre-commercialization. First, lack of facilities and financing to move the research to the concluding stages. Second, we found situations where significant research results had been collected, with evidence of possible utility of the process and product, but no demand by the end-users.[7] Third, failure to commercialize sometimes resulting from institutional rigidity much of which relates to the ways traditional PRIs and universities are set up. There are two issues that recurred in our interviews namely, who initiates the process (the PRI or a firm/ entrepreneur?); and what form of formal or informal contract guides the process?

Econometric analysis on the factors that enable foreign and local collaborations based on survey data (not reported here) shows that human capital affects positively and significantly both foreign and local collaboration. In other words, the more the presence of staff in local organizations with Masters' and Doctoral degrees, the greater the incidences of collaboration. The econometric analysis also reveals that appropriate government innovation incentives affect both types of collaboration positively and significantly.

5.5 The role of policy and the state in enabling a Sectoral System of Innovation (SSI) for agricultural biotechnology

While it is inevitable that actors and organizations internalize the inefficient incentive structures into their day-to-day practices and general attitudes over a period of time, it is the role of institutional frameworks to gradually introduce newer incentives that could bring about attitudinal shifts over a period of time. Policies both direct and indirect play a pivotal role on what the intrinsic absorptive capacity of the local actors is, and on how they choose to interact and collaborate. Intangible factors especially attitudinal in nature, determine application of knowledge to innovative activities in local contexts in late-comer countries. The social practices and habits of scientists may well be as important as the conducive physical and scientific environment for inventive activities.

5.5.1 Policies and institutions for agricultural biotechnology

The *national biotechnology policy* document states that the Federal Government of Nigeria 'supports biotechnology because of its immense potential to more rapidly contribute to sustainable food security and economic growth'. The government established the National Biotechnology Development Agency (NABDA), and approved the National Biosafety Guidelines in 2001. NABDA was established by the Ministry of Science and Technology to promote the development of biotechnology in Nigeria. The agency has the mandate of creating awareness and coordinating activities related to biotechnology.

The National Focal Point, which comprises the Federal Ministry of Environment, the National Biosafety Authority (NBA),[8] the National Biosafety Committee (NBC), the National Biosafety Technical Sub-Committee, the Institutional Biosafety Committees, the National Biotechnology Development Agency (NBDA), has developed a *National Biosafety Framework* (NBF) to provide guidance on the implementation of Nigeria's biotechnology programme. This framework is a combination of policy, legal, administrative, and technical instruments that will regulate all biotechnological work to minimize or eliminate any potential hazards. It is also intended to ensure the safe transfer, handling, and use of biotech materials that may have adverse effects on the conservation and sustainable use of biological diversity, taking into account risks to human health. The framework is meant to provide a one-stop clearing-house and has established a 15-man NBC to oversee the implementation of the National Biotechnology Programme in 2000. The committee has developed a *draft Biosafety Bill*, which was yet to be circulated for national debate in 2006. The draft Bill generally portrays products of biotechnology as safe for animal and human consumption, it, however, advocates strict adherence to the 'precautionary principle' and mandatory labelling of all products of agricultural biotechnology to protect 'consumers right to know'. The National Focal Point is responsible for all correspondences with importers, exporters, and applicants on movement of products of modern biotechnology. Pending the passage of the National Biosafety Bill, the Ministry of Environment is proposing the establishment of an independent NBA. The NBC serves as the Competent National Authority for biosafety in Nigeria. The committee is responsible for the safe management of biotechnology activities, including research, development, introduction, and the use of GM organisms.

The NBC has also established National Biosafety Technical Sub-committees (NBTS) to focus on sectoral interests such as agriculture, health, industry, and the environment. The sub-committees review proposals for research and recommend the conditions under which experiments should be conducted. They are to provide technical advice to the NBC and contribute to its functions in relation to contained use, field trials, release and placement on the market. The framework also requires the establishment of Institutional Biosafety Committees (IBC) by all institutions in Nigeria, both private and public (e.g. research institutes, universities, international research centres, etc.), which plan to undertake biotechnology research and/or development. The IBC shall consult and seek approvals from the NBC and implement recommendations from NBC. Despite all these developments, at the time of the survey in 2006, Nigeria had no laws governing modern agricultural biotechnology and biosafety.

Despite the National Biosafety Guidelines that were approved in 2001 with a provision for field-testing bio-engineered crops, no bio-engineered crop variety had been approved for field-testing as of 2006. Field trials of transgenic crop, animals, and material could not be brought into the country and no law existed to approve biotechnology crops for food, processing, and feed. An application was pending before the NBC to introduce the first LMO virus-resistant cassava variety for field trials in Nigeria. The variety was developed in the US in collaboration with two Nigerian scientists. The processing of the application for field trials in Nigeria was delayed because the NBC says it lacks funding to meet and deliberate on the application. If the application were approved, it would allow local regulators and scientists to gain familiarity with biotech crops and encourage development of workable biosafety systems.

The NBC has also drafted another Bill on Nigerian Biotechnology Programmes, which has not been passed into law by the National Assembly. Even public debates on the draft Bill had not commenced at the time of the survey and this uncertainty and lack of policy impetus keeps agricultural biotechnology activity low in the country.

On the question of public attitudes towards biotechnology and GM products, the results of a focus group survey conducted within the country on the attitude of the public to biotechnology revealed that about 40% of respondents would not mind consuming bio-engineered food products. Many respondents especially among those with little education were ignorant of biotechnology and its potential usefulness. While respondents did express concern about the long-term health effects of consuming such products (which were also not well-informed), these concerns seem to be overshadowed by their basic need for affordable food. The survey also revealed a marked preference for biotech products develop locally to those that are imported.

However, a closer review of all the policy developments in conjunction with biotechnology reveals the following:

(a) Nigeria's biotechnology programme is fragmented and the linkages between the national biotechnology policy, biosafety guidelines, draft biosafety Bill, and its larger STI policy are all unclear.
(b) While policymakers seem very aware that biotechnology relies on strong public science and on the availability of human capital and funding, very little effort seems to be ongoing to ensure that the structural constraints of the innovation systems at large that also affect the SSI for agricultural biotechnology are addressed.
(c) There are no policy efforts directed at some of the most serious deficiencies in the SSI for agricultural biotechnology, such as the absence of ancillary institutions to help focus on product development and incentives that aim to improve collaboration.

5.5.2 Programmes and organizations for agricultural biotechnology

Several Nigerian organizations have the mandate of performing various biotechnology related functions including research, such as NNMDA, NABDA, NACRAB, and NABP as were discussed in the earlier section. Table 5.5 contains a list of the biotechnology programmes and the targets set therein.

Table 5.5   Biotechnology programmes in Nigeria

Biotechnology programmes	Major goals
1. National Biotechnology Development Agency (NABD), Abuja.	Development of appropriate biotechnology policies and programmes
2. SHESTCO National Biotechnology Advanced Laboratory, Abuja.	Central laboratory for advance biotechnology research in Nigeria.
3. National Center for Genetic Resources and Biotechnology (NACRAB), Ibadan	The use of biotechnology in conservation of Nigerian genetic resources
4. National Biosafety Frameworks, Federal Ministry of Environment, Abuja	Capacity building and effective regulation of GM products in Nigeria.
5. Nigeria Agriculture and Biotechnolgy Project (NABP), IITA, Ibadan.	Promote the adoption of biotechnology for enhanced agricultural productivity

Source: Kuta (2004).

But again, these are not well-coordinated and our survey and interviews suggest that the organizations are ineffective in performing their functions largely because of over-lapping and incoherent mandates. As the discussion in the previous section reveals, most new organizations set up for performing biotechnology-related functions have no clear mandates, no funding and have to deal with the ambiguities of operating within an uncertain policy framework. This exacerbates the inefficiency of the organizational structures that are already over-burdened by staff limitations and low morale of actors. The institutional inertia thus gets internalized and is an issue of equal importance for both new organizations as well as old ones that have existed for several decades, in the context of late development.

Rectifying this and providing a more robust basis for interactive learning through organisational collaborations in the SSI for agricultural biotechnology will require policies that not only set out a set of objectives (as in the National Biotechnology Policy and the Biosafety Framework) but also targets the elimination of basic hurdles to innovation. These include:

(a) *Physical infrastructure and networking*: Autonomous telecommunications faculty, such as telephone, fax and Internet in the universities falls largely below what may be required for good research and teaching work, getting to know and linking up with similar programmes with other universities and institutions within the country or the rest of the world and for dissemination of research efforts. There is a critical shortage of power and water; a situation that calls into question the commitment of governments on research investment.

(b) *Knowledge infrastructure*: High quality research requires equipment and investment in sector specific facilities and laboratories at the national and regional levels. Mostly research equipment are not available, and even laboratories created to facilitate them depend on the availability of foreign grants for their survival. Basic chemicals and reagents are hard to find and graduate students often have to rely on own funding to carry out experiments. Researchers and their initiatives are stunted by lack of funding and the resulting intellectual isolation gets internalized into informal norms and codes of conduct.

(c) *General and specific innovation incentives*: Sectoral policies are important, coordination between sectoral policies, especially those that link research in traditional sectors and new technologies are critical. The comparison between the two countries shows failures (albeit of different kinds) on linking their investments into biotech R&D (in the public sector) with the enterprise sector. There are no specific incentives created in the local innovation systems that promote commercializing of research results. Newer and more dynamic policies and incentives are required to bridge the gap between research and production (private sector) activities in these countries.

5.6 Case study: The cassava success in Nigeria

Four key factors drove the collaborative success of production expansion of cassava in Nigeria. First, the application of scientific and contribution institutional support of the International Institute of Tropical Agriculture (IITA)'s new high-yielding Tropical Manioc Selection (TMS) varieties boosted cassava yield by 40% without fertilizer application (Bokanga and Otoo, 1991; IITA, 1991). Second, higher consumer demand for cassava by rural and urban households encouraged the farmers to plant more land to cassava. Third, the inventive participation of Nigerian fabricators, who fashioned a new mechanical grater to prepare *gaari* to reduce labour, especially female labour, from processing for planting more cassava. Fourth, regional collaboration through the Africa-wide biological control programme was highly influential in averting the devastating cassava mealy bug epidemic.

In the mid-1980s, the Nigerian government invested in measures to diffuse the improved TMS varieties among the farmers. By the late 1980s, the TMS diffusion in Nigeria had become an Africa's agricultural success story. In 1989, IITA researchers conducting the Collaborative Study of Cassava in Africa (COSCA) found that farmers in 60% of the surveyed villages planted the TMS varieties, mainly boosted by the fact that the TMS varieties were ideal for *gaari* preparation. Cassava production in 2003 was about 34 million metric tonnes from a total land area of about 3.1 million hectares, giving an average yield of about 11 tonnes per hectare (RMRDC, 2004).

From the mid-1980s to the early 1990s, the diffusion of the TMS varieties was rapid and as of mid-1990s, 60% of the area cropped with cassava in Nigeria is planted with improved varieties (IITA, 2004). As a result, cassava production per capita increased significantly. However, the demand for cassava did not rise significantly leading to a decline in the price of cassava failing to contribute to increase in the real income of the rural farmers that produced them, thus penalizing urban households who consume cassava as their staple food. These temporary shocks notwithstanding, Nigeria's production of cassava was at 40 million tonnes in 2005 and the projected target for 2020 is set at 60 million tonnes.

IITA is highly active in creating more improved varieties of cassava and trials of 40 new varieties of cassava are continuing, with the aim of replacing improved varieties that are not resistant to cassava mosaic disease. Other local organizations that have contributed to the development and improvement of cassava include the National Root Crops Research Institute (NRCRI), a Federal Government Institute under the Federal Ministry of Agriculture and Water Resources with the main mandate of conducting research into the genetic improvement, production, processing, storage utilization, and marketing of root and tuber crops of economic importance (cassava, yam, potato, sweet potato, cocoyam, ginger) and other root crops. The Root and Tuber Expansion Programme which is an initiative of the Federal Government of Nigeria, is also involved in promoting accelerated expansion of cassava production in Nigeria.

Ensuing the dip in demand that cassava farmers experienced in the 1990s, the Nigerian state began to engage proactively through policies that promote the production and use of the newly developed varieties and presently five varieties have been released, which are rapidly being distributed for cultivation through the Cassava Growers Association of Nigeria that was formed as a distribution channel.

Tables 5.6 and 5.7 present the results of a study carried out on the cassava sub-system and the collaborative patterns of interactions observed therein in 2008 by National Agency for Science and Engineering Infrastructure (NASENI). As part of the survey, a range of actors (see Table 5.6) were inter-viewed and a total number of 127 questionnaires were retrieved. Among

Table 5.6   Actors configuration, activities, and constraints

Actors	Activities	Organizations included in this study	Constraints
Farmers	Produce crop for food and industrial use; preserve genetic resources	Mostly rural small-scale farmers. *Lagos State: 34 farmers located in 9 different local government area namely:*	Resource-poverty, crop pests and diseases, unpredictable markets.
Scientists (e.g. researchers, educators, agric. extension officers in Universities and Public Research Institutes)	R&D to improve crop varieties; provide solutions to farming constraints e.g. diseases and pests, develop processing technologies	**3 Universities:** *Kwara State (27 Qs)*: Univ. of Ilorin, Ilorin *Osun State (36 Qs)*: Obafemi Awolowo Univ., Ile-Ife *Ogun State (28 Qs)*: University of Agriculture, Abeokuta **2 Research Institutes:** Agricultural and Rural Management Training Institute (ARMTI), Ilorin (10Qs) National Centre for Agricultural Mechanization (NCAM) Ilorin (10 Qs)	Bureaucracy, limited funding, inadequate research facilities and poor infrastructure
Private entrepreneurs and investors	Set up industries for processing and fabrication of machineries and equipments. Bring cassava to the markets, process cassava, diversify products and export.	Micro, small, medium and large scale enterprises; and industries. *Lagos State: 32 Producers in 9LGAs 35 Fabricators in 8LGAs* **A Total number of 67 Qs**	Restricted capital markets; high interest rates, poor infrastructure, unfavourable external market conditions.

Source: NASENI Empirical Survey (2008).

the several results of the study, the ones most pertinent to us related to the collaborative patterns that led to the development of cassava varieties and their use in production, as well as the main drivers of this change.

Table 5.6 illustrates the different actors and communities spanning across scientific, industrial and marketing, which interrelate through their respective functions to ensure effective innovation. While farmers may be seen as the critical actors in the direct production of cassava, the cassava success story points to two major drivers:

(a) The presence of a sustained research programme in a PRI (in this case IITA) that was able to drive the cassava research agenda throughout the local scientific and research community within the country.
(b) A process technology-oriented innovation (for the production of *gaari*) that matched the scientific innovation of the TMS varieties of cassava, that led to its increased demand and widespread uptake for production among local farmers.

This is further amplified when the main areas of work focus among all actors involved in cassava sub-system are compared. As Table 5.7 shows, the majority of actors engaged in cassava production are involved in research and teaching (67.7%) while about 28.3% of the respondents are involved in various kinds of consultancy, 18.1% are involved in production of cassava, 4.7% participate in testing and laboratory services while another 7.1% are involved in marketing activities and 6% are engaged in other activities such as extension services (Table 5.6 shows the nature of the organizations).

Furthermore, the NASENI survey revealed that specific details of cassava research that respondents are involved in can be categorized in five broad areas:

Table 5.7   Area of focus of cassava actors

Nature	Frequency	Percentage
Research	86	67.7
Teaching	80	63.0
Consultancy	36	28.3
Production	23	18.1
Testing & lab. services	6	4.7
Marketing	9	7.1
Others (extension)	5	4.0

Source: NASENI Empirical Survey (2008).

(i) Research directed towards *extension education* on improved varieties (17.2% of respondents).

(ii) *Improvement and production of cassava crop* using applicable cutting edge *technology* (modern machineries or farm mechanization) (42.9% of respondents).

(iii) *Livestock management and veterinary activities* (20.2%).

(iv) Evaluation of *nutritive value of cassava*, which include studies such as the adverse effect of hydrocyanic acid in cassava on humans (11.1% of respondents).

(v) *Processing and disease control* (9.3% of scientists).

The cassava success was catalysed by collaboration efforts of all the actors across the scientific, industrial, marketing, and use domains, which was only possible through coordination of the various agencies and organizations involved in various stages of the process. What it shows that is particularly relevant to the analysis conducted in this chapter is that although the state followed in strengthening the ongoing efforts and in promoting demand in the ex-post stages, especially since the late 1990s, state has not been proactive in tele-guiding activities and capabilities of relevance to local needs and food security in agriculture. The state's role in scanning the horizon for opportunities, especially those brought on by agricultural biotechnology, and utilizing them to meet the local challenge, is yet to be realized in Nigeria's case.

This finding is reinforced through other work we have done in Nigeria, for example, in the case of biopharmaceuticals and health wherein the state once again reacted to a successful collaboration between a Nigerian PRI and local healers that led to the development of a sickle-cell drug (NIPRISAN), instead of carrying the baton of innovation itself (see Oyeyinka and Gehl Sampath, 2007; Gehl Sampath, 2009).

5.7 Summing up

Nigeria's sectoral system for agricultural biotechnology is fragmented, with the public sector institutions being the primary actors. The SSI is unable to perform as a result of several major resource constraints including: the lack of human capital, physical and scientific infrastructure, and the lack of ancillary institutions to promote product development. The analysis conducted in the foregoing sections shows that these are a result of fragmented policymaking for the sector, on the one hand, and the state of the national innovation system at large, which also affects sectoral innovation activity in agricultural biotechnology.

In the absence of additional policy measures that aim at mitigating the imperfections of the innovation system at large, and those that cater to the specific features of biotechnology-based research and entrepreneurship, the sectoral system of innovation for biotechnology is likely to remain stunted.

Following the discussion on the role of the state from Chapter 2, there are two primary ways in states could provide a conducive institutional framework for innovative activity. First, are there resources and personnel put in place to match the goals of the programmes and policies set out in various documents. Second, is the state able to coordinate these activities and rally actors, through various incentives, towards achieving the goals. While the first calls only for the allocation of resources in a holistic way, the second demands a much more proactive stance, where the state champions innovative activities in a particular sector, in this case agricultural biotechnology. The state, in the Nigerian case, does not seem to be in a position to perform either of the two functions in the context of agricultural biotechnology. The cassava case highlights the lethargy and lack of direction in state's efforts, which is more reactionary than that of express leadership, where STI investments are viewed as strategic inputs to resolve pressing local challenges.

The cassava case, an example of sporadic success discussed in this chapter helps to underscore the analysis of the main shortcomings of the agricultural biotechnology systems of innovation. It shows that success was determined by the presence of a sustained research and science base that was devoted to the issue, and concomitant innovations on the marketing and commercial side, that promote the uptake of the scientific innovations, which calls for systemic coordination and functionability across a wide range of capabilities. It points attention to the ability of the state to enact policies and incentives that promote coordination and collaboration, and the presence of a vast public science base to enable biotechnology-based innovation.

6
Kenya's Incipient Innovation Capacity in Biotechnology

6.1 Introduction

The development of biotechnology in Kenya over the last two decades reflects a steady transition from traditional, low-end biotechnologies such as fermentation, bio-fertilizers, and tissue culture techniques (Odame et al., 2003) towards more sophisticated, modern techniques and applications comprising the use of molecular markers, novel vaccines, diagnostic tools, and genetic engineering.

The initial developments in biotechnology in Kenya can be traced to the early 1980s with the application of tissue culture in crops such as citrus fruits (Kenya Agriculture Research Institute, KARI) and pyrethrum (University of Nairobi). In many organizations, the application of these techniques was largely done as an increment to ongoing conventional breeding programmes. By the end of the 1990s, several local research organizations had taken up tissue culture and were applying it across a broad range of products. Table 6.1 below shows the extent of application of tissue culture a decade ago, by organizations and products.

During the same period, the use of biotechnology in livestock research and development also began, mainly focusing on the generation of disease diagnostic technologies employing hybridoma and DNA molecular techniques. KARI pioneered the application of molecular marker selection techniques in 1995, focusing on maize breeding geared towards isolating and developing cultivars resistant to insect pests, maize streak virus, and drought tolerance. Since then, molecular marker assisted breeding has been applied in the following areas in the country (Gichuki, 2006):

- Characterization and mapping of Grey Leafy Spot (GLS) resistance genes using microsatellite markers.

141

Table 6.1   The state of tissue culture in Kenya as of 1998

Institution	Crops
KARI	Pyrethrum, banana, strawberry, cassava, potato, and sweet potato
KEFRI	Camphor wood (*Ocotea usambarensis*), Silky oaks (*Gravillea robusta*) Mvule (*Chlorophora excelsa*) and *Eucalyptus grandis*
Universities	Banana, citrus fruits, sugarcane, pawpaw
*Oserian Dev. Company	Flowers
*Genetic Technology Limited (GTL)	Bananas, sugarcane, flowers
Kenya Seed Company	Vegetables
Tea and Coffee Research Foundations	Tea and coffee

Source: Wafula (1999).
Note: * Private companies.

- Introgression of MSV resistance into maize lines resistant to GLS disease.
- Selection of smut resistance in sugarcane.
- Diversity studies for sweet potato and cassava.
- Characterization of indigenous species of cattle and forages.
- Characterization of tsetse flies.

The first modern biotechnology-based product to be developed in Kenya was a genetically modified (GM), virus-resistant (VR) sweet potato. This project started in 1991 and was a public–private partnership (PPP) between the United States Agency for International Development (USAID), KARI, and the Monsanto Company. Although the product itself did not stand the rigour of field trials, the GM sweet potato was held up to be an example of successful collaboration between the global private sector and a latecomer public research institution due to three key reasons:

(a) Its capacity building element, where several KARI scientists were trained in Monsanto laboratories during early stages of the project.
(b) Monsanto and KARI signed a non-exclusive, royalty free licensing agreement in 1998 which allowed KARI to develop other transgenic virus technologies for sweet potato building further on the existing work.
(c) KARI is permitted to protect the new creations under Kenya's plant breeders' protection regime.

Table 6.2 elaborates upon the collaborators and partners in the sweet potato project.

Since then, work on several other genetically modified crops has been ongoing in Kenya through the public-private partnership mechanisms (Kirea, Awuor, and Asali, 2003; Clark et al., 2005). Table 6.3 shows the current status of modern biotechnology in Kenya.

Table 6.2 Collaborators and partners in the sweet potato project

Collaborators	Nature of collaboration
KARI	• KARI scientists working at Monsanto Company were involved in the development of the gene constructs, transformation protocols, and regeneration systems for the transgenic sweet potato. • KARI staff members carried mock trials on CPT 560 and other local varieties. • KARI has developed an operational biotechnology laboratory for further transformation of local African sweet potato genotypes.
USAID/Agricultural Biotechnology Support Programme (ABSP)	Provided the financial assistance
Monsanto Company (St. Louis, US)	• Donated the genes of interest. • Monsanto also supported the initial research support for the genetic transformation of six Kenyan sweet potato varieties.
International Potato Centre (CIP)	International Potato Centre researchers collected data on crop establishment, crop vigour, vine, and storage root yields and response to sweet potato virus based on protocols during the trials.
KEPHIS	Granted KARI a biosafety permit for on-station field-testing after its approval by National Biosafety Committee.
ISAAA	• Material transfer agreement • Brokered the Intellectual Property Rights (IPR) negotiations between Monsanto and KARI and a royalty-free licence agreement between the two was signed in 1998. • Helped in identification of appropriate partners for the different implementation stages
Danforth Plant Science Centre, US	Offered technical support

Source: Authors' survey (2006–7).

Table 6.3 Status of modern biotechnology (genetic modification) projects in Kenya, 2006

Transgenic crop/product	Desired trait	Institutions involved	Year of approval	Status, 2006
Bt Maize	Insect resistance	KARI/CIMMYT with financial support from Syngenta Foundation	2001 (leaves) 2003 (seeds)	Undergoing contained field trials/ evaluation since May 2005
Bt Cotton	Insect resistance	KARI/ MONSANTO	2003	Contained trials
Transgenic sweet potato	Viral resistance	KARI/ MONSANTO/ Danforth Centre (US); ARC-VOPI; ISAAA	1998	Contained trials
GM Cassava	Cassava mosaic virus	KARI/Danforth Centre (US); USAID (ABSPII)	2003	Contained trials
Rinderpest vaccine	Disease control	KARI, Pirbright (UK), University of California	1995	Contained trials

Source: Modified by authors from IRMA (2004) and Gichuki (2006).

6.2 Key actors generating agricultural biotech innovation and knowledge

The majority of research projects on agricultural biotechnology in Kenya are concentrated in the public sector with a predominance of donor funding. Of these, KARI is involved in the largest number of biotechnology research programmes and projects spanning both crop and livestock domains. Most of the modern biotechnology research projects are collaborative projects (through private–public partnership mechanisms) between KARI and international partners with financial support from the international private sector and international donors (Clark et al., 2005). Table 7.4 shows the range of research and product development activities involving Kenyan scientists and researchers as of 2006. The current agricultural biotechnology projects are as listed in Table 6.4 below.

Table 6.4 Current agricultural biotechnology projects in Kenya (2006)

Technology	Developed by	Description	Year	Status
Transgenic sweet potato CPT 560 With resistance top sweet potato feathery mottle virus	Wambugu, F. and Monsanto USA	Sweet potato transgenic lines (CPT) expressing coat protein gene of sweet potato feathery mottle virus (American isolate)	1996	Development discontinued in 2002
Transgenic sweet potato CPT 560 & KSP 36 with resistance to sweet potato feathery mottle virus	S. T. Gichuki, J. Machuka, E. M. Ateka, I. Njagi, J. Irungu	Sweet potato transgenic lines (CPT 500, KSP36) expressing coat protein and replicase genes of sweet potato feathery mottle virus (Kenyan isolate)	2003–5	Still under evaluation at contained biosafety glasshouse level
Transgenic maize with resistance to stem borers	CIMMYT and KARI scientists	Transgenic maize events carrying Cry IAb and Cry IBa genes (isolated *from Bacillus thuringiesis*) for protection of maize against stem borers	2001–5	Undergoing confined field evaluation and conversion
Transgenic maize (MON 810)	E. M. Ateka, H. Ngesa, S. T. Gichuki and Monsanto South Africa Ltd.	A maize hybrid (DKC8073YG) carrying a cry IAb gene (isolated from Bacillus thuringiensis) for protection of maize against stem borers	2005	Application for evaluation under containment has been made to the NBC and waiting approval
Regeneration and transfor-mation protocols for tropi-cal maize	Songa J. Binot J. Sitieney J.	Regeneration of Kenya maize varieties through embryo rescue techniques	2004	Several KARI maize lines have been successfully regenerated and transformed
Bt isolates with potential for controlling Larger Grain Borer (LGB) and Maize Weevil (MW)	Songa J. Wamaitha J.	Local Bt. isolates that have been bio-assayed and have shown potential for controlling the larger grain borer and maize weevil	2004	Characterization of the isolates continuing. Search for more isolates continuing. Genes will be cloned from the best isolates

(Continued)

Table 6.4 (Continued)

Technology	Developed by	Description	Year	Status
Transgenic cassava with resistance to cassava mosaic disease (CMD)	Taylor N. Faquet C. Gichuki S.T. Njagi I. Wangai A.	Cassava expressing the AC1 replicase and the DI defective interfering particle gene of cassava mosaic virus	2002–5	Confined field trials proposed. New products of Kenya varieties being developed
Cloning of two genes fro starch branching enzyme IIa(*sbella*) and IIb (*sbellb*) in sorghum	Joel Mutisya	Both genomic and cDNA sequences of *sbella* and *sbellb* were cloned. CDNA sequences were cloned from KARI Mtama 1 sorghum variety. Thyme are between 2.6 to 2.7 kb in size	2003	The *sbellb* sequences were deposited in the gene bank but *sbella* has not yet been deposited
A promoter from sorghum to drive gene expression	Joel Mutisya	Derived from sorghum genome. Used to drive a gene for starch branching enzyme IIb *in vivo*. Size is 2.7 kb. Fully characterized. Ideal for gene expressions in seeds and cereal crops.	2003	Sequence has been deposited in the gene bank. It is being used to make constructs to improve starch quality in sorghum.
Gene transformation construct for sorghum starch	Joel Mutisya Mercy Mbogori	Contains a gene for starch branching enzyme in antisence orientation fused with *gfp* as a reporter and 35S as promoter	2005	Cloned in a binary vector and stored for use in appropriate agro- bacterium strain. Its ready for transformation work

Source: Gichuki (2006).

Besides KARI, there are a number of other national institutions (private and public) as well as international public institutions involved in biotechnology research and development in Kenya.

Some of the other actors in the agricultural biotechnology innovation system that were identified in Chapter 2 were also present, such as farmers, firms providing inputs (such as seeds, pesticides, fertilizers, animal feed energy) or services such as transport, agricultural machinery rentals, credit, insurance, animal health, etc.) to farmers, agro-processing firms, retailers, supermarkets, commodity boards, training, research and development institutes and tertiary colleges, universities and agricultural extension and training services, ministries of agriculture, trade and industry, environment, health and standards, and regulatory and quality control institutions. However, as in the case of Nigeria, Kenya does not have any local firms that are involved in development of biotechnology-based products. Viewed across the five domains that we presented in Chapter 2, they could be schematically represented as below (Figure 6.1).

6.3 Science, technology, and innovation investments

As early as 1990, the National Committee on Biotechnology Advances and its Applications (NACBAA) report recommended the need for strengthening the country's scientific, legal, and bureaucratic capacities in order to harness the benefits of biotechnology. Wafula and Falconi (1998) estimated that by 1996, there were only 56 scientists involved in biotechnology research activities in Kenya. These scientists accounted for 80% of biotechnology research in Kenya while the remaining 20% was conducted by scientists in international organizations in Kenya (Odame 2005). Our survey confirms the result that were arrived upon by earlier studies that even though the majority of Kenyan scientists may have basic scientific knowledge on issues of genetics and molecular biology, only a few of them are specialized enough to conduct research and development activities in what was identified as biotechnology-based work in Chapter 2.

6.3.1 Human resources

Our survey also found that the limited capacity that is being created is focused more on tangible infrastructure (such as labs and equipment) and is not matched by the expansion of human skills to utilize these facilities as part of structured research agendas for the Gene Revolution. This once again confirms earlier results on Kenyan biotechnology

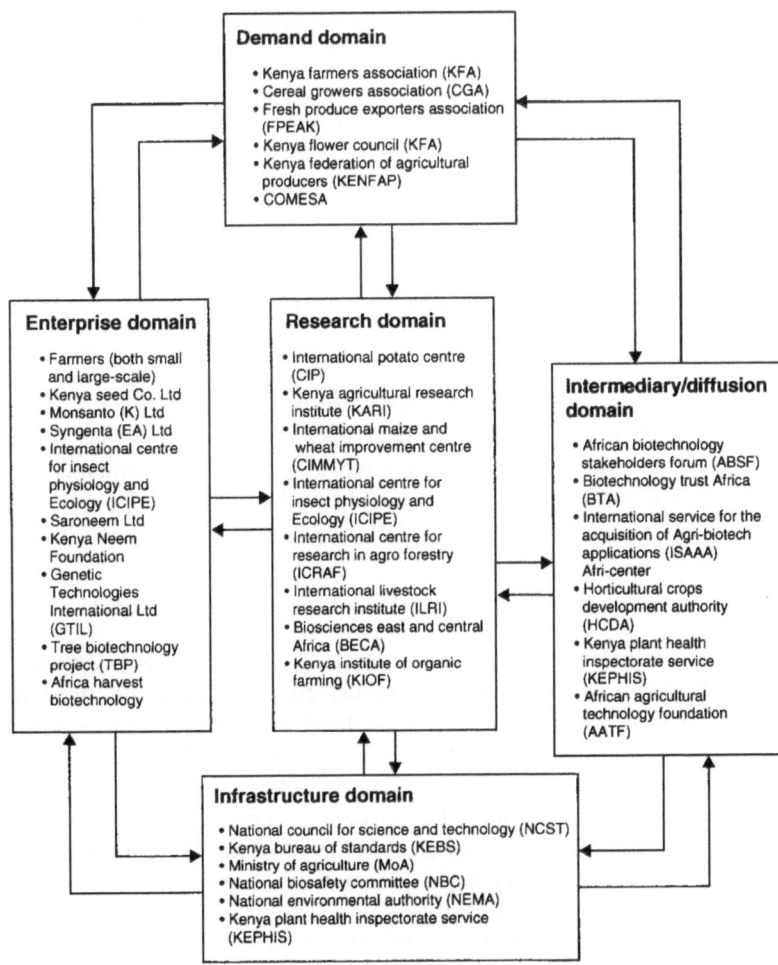

Figure 6.1   Components of the agricultural biotechnology innovation system in Kenya
Source: Authors.

capacity (see for example the survey by Odame and Mbote, 2000) which concluded that over the years capacity-building has focused more on hardware components (expanding the physical facilities) and postgraduate training at MSc and PhD levels. This over-emphasis on hardware components has resulted in an increase on the demand for non-scientific

Table 6.5 School enrolment ratios, Kenya 2000–5

	2000	2001	2002	2003	2004	2005
School enrolment, primary (% gross)	98	–	94	111	111	–
School enrolment, primary (% net)	67	–	63	77	76	–
School enrolment, secondary (% gross)	39	–	41	44	48	–
School enrolment, secondary (% net)	33	34	35	37	40	–
School enrolment, tertiary (% net)	3	3	3	–	3	–

Source: World Development Indicators (2007).

staff to manage the expanded physical facilities, thus explaining the low scientific – non-scientific staff ratios captured in our survey.

Table 6.5 shows once again the low rates of continuation among those enrolled from primary to secondary to tertiary levels of education. Our survey also found that the share of staff with Phd degrees in the public research institutes was extremely small, again confirming the findings that most of the 'core research' staff required to conduct biotechnology-based R&D are missing from the expanding infrastructure endowments.[1] And, as in the case of Nigeria, we found that most of the biotechnology work was again concentrated around tissue culture and other basic biotechnologies, rather than seeing pockets of excellence across a broader range of expertise, given the broad range of international collaborations going on in the country.[2]

The government has sought to address the issue of adequate human and technical capacity by establishing courses in biotechnology in most of the public universities in Kenya. In fact all the six public universities across the country are offering biotechnology courses at undergraduate and post-graduate levels. For example, Kenyatta University offers both BSc and MSc courses in Biotechnology, while Moi University's School of Agriculture and Biotechnology has a BSc course in Agricultural Biotechnology. The University of Nairobi in 2005 established the Center for Biotechnology and Bioinformatics (CEBIB) as a centre of excellence to facilitate capacity-building and generate marketable products by harnessing biotechnology. CEBIB's mandate is to enhance knowledge and skills in biotechnology and bioinformatics to impact on agricultural and industrial output, health, and environmental management.

CEBIB's formation underscores the realization that biotechnology is an interdisciplinary subject with wide ranging applications of scientific and engineering principles in different fields such as agriculture, food and feed, medicine, industry and the environment, which are of profound importance to mankind. The centre has the following key objectives:

1. To strengthen national capabilities in the field of basic sciences and technology and in the development of research in biotechnology and bioinformatics.
2. To promote and conduct basic research in the areas of molecular biology, biotechnology and bioinformatics.
3. To facilitate the application of biotechnology in research and encourage its use for the development of marketable products.
4. To offer training facilities for manpower development in biotechnology and bio-informatics at the national and regional level.
5. To institutionalize links between universities, scientific research institutions, and the private sector.
6. To network with institutions in developed and developing countries as well as the international centres of biotechnology and bioinformatics.

It is expected that the integration of biotechnology courses within the university curricula as well the emergence of training centres such as CEBIB will help boost the country's human capacity in biotechnology, and more specifically, modern biotechnology in the years to come. The bigger challenge, however, is to equip these centres with the necessary training materials and facilities, on the one hand, and relevant manpower, on the other. Considering that Kenya currently allocates less than 0.5% of its GDP to science and technology, this feat may be difficult to achieve if budgetary allocation to science and technology is not increased. It also calls for universities and research institutes to seek alternative, innovative funding mechanisms, the incentives for which are presently not in place.

6.3.2   Funding

The National Council on Science and Technology in Kenya has a broad mandate of focusing on agricultural innovation and new technologies of importance to the country, such as biotechnology and ICTs. Despite this, the amount of resources that are allocated to public-sector organizations for research is negligible. Organizations such as the Kenya Industrial Research and Development Institute (KIDRI) and KARI which have the mandate to develop technologies for the use of local entrepreneurs in both traditional and new technology sectors operate with extreme staffing and funding shortages, and hence are not able to fulfil their mandates even partially. Most researchers at universities and PRIs complain regularly of a lack of funds and initiative on part of the government to support and direct relevant research. The extraordinary reliance on external, donor funding for research, which is at best sporadic and not dependable, means that innovative activities in academic institutions in the country continue at a rate that hardly reflects its true potential.

6.4 Interactive learning

The survey sought to identify the following key issues in the context of agricultural biotechnology system of innovation in Kenya:

- The type/nature of collaborations, for example research, financing, marketing, or distribution?
- Who the key partners and collaborators are (both national and international; private or public)?
- How the linkages were initiated and established, and the factors that play a role?
- The structure and intensity of the agreements with partners/collaborators, for example intellectual property clauses, capacity-building elements among others?

Specifically, we were engaged with questions that stand out when one analyses biotechnology developments in the country. For example, why is KARI's example of the sweet potato, which was regarded as an example of a fruitful collaboration in agricultural biotechnology, not resulted in other variations of the product that are locally suited despite the training that the scientists received as part of the initial phase of the project? Why do we not observe more such collaborations that expressly target innovation capacity for agricultural biotechnology in the country? What is the broader impact of the international collaborations presently ongoing in the country, how do they help in improving Kenya's intrinsic capacity to conduct biotechnology innovation? Our survey found the following factors instrumental in limiting Kenya's capacity to engage and expand in agricultural biotechnology.

6.4.1 Lack of knowledge infrastructure

A review of previous studies all agree on this point: Kenya still has a shortfall of adequate laboratory capacity and facilities needed to effectively exploit biotechnology (Wafula and Falconi (1998); Odame and Mbote (2000); Odame, Kameri-Mbote and Wafula (2004), and Quemada (2002) to mention a few). Human skill, the other component of knowledge infrastructure, is highly limited, a point that has been

Table 6.6　Financial support to entrepreneurial activity, Kenya

Financial Support	2000	2001	2002	2003	2004	2005
Domestic credit to private sector (% of GDP)	28	25	26	24	24	27
Interest rate spread (lending rate minus deposit rate)	14	13	13	12	10	8
Market capitalization of listed companies (% of GDP)	10	8	11	28	24	36

Source: World Development Indicators database.

discussed at length in the previous section. This calls for the need for continuous training in the fields of modern biotechnology and genetic engineering. While local universities have begun to respond to this challenge by designing relevant courses as well as establishing specialized centres to handle capacity development in biotechnology, there is a need to ensure that the courses are relevant, and the manpower being generated is of the kind needed by the private sector and the research institutes. There is also a need for a dynamic biotechnology policy to promote entrepreneurial efforts by academic and research institutes, as well as create a new culture of collaboration between public research and enterprise. Our survey found that the research institutes and their research programmes are not well received by the enterprise sector, and there is very little reliance on the services provided by them in firm-level strategies.

There are inherent weaknesses in several institutions that are fundamental to the creation of new knowledge and the use of already existing knowledge in innovation activities. As Table 6.6, for example, shows financial support institutions to promote local innovation and entrepreneurship in Kenya, which have not been performing well and have been constantly on the decline since 2000. Domestic credit to private sector has been on the decline between 2000 and 2005 and other macro-economic indicators such as net inflows as percentage of GDP (see Table 6.6) show that investing in the economy is not a highly profitable activity. R&D investments in the economy have similarly been very low between 2000 and 2005, further exemplifying the weak institutional environment for innovation (see Table 6.7).

6.4.2　International collaborations and innovation capacity

Although Kenyan PRIs are actively engaged in several projects that involve more demanding biotechnologies such as GM technologies

Table 6.7 Investment and R&D

Investment and R&D	2000	2001	2002	2003	2004	2005
Foreign direct investment, net inflows (% of GDP)	1	0	0	1	0	..
Merchandise imports (current US$)*	3,104.99	3,192.00	3,244.83	3,725.29	4,552.73	6,360.00
Research and development expenditure (% of GDP)	..	..	..	..	..	..
Researchers in R&D (per million people)	..	..	..	..	..	..

*Amounts in 100,000.
Source: World Development Indicators database.

(see Table 6.1), none of these projects has led to the commercial cultivation of GM crops within the country and most of the products were
undergoing contained trials at the time of our survey.

More importantly, these projects have also not contributed to building local research capacity in significant ways because of the observed
tendency of international private-sector companies to bring into the
country finished (already modified) products for trials, thereby limiting the active participation of local public-sector institutions and their
researchers in the research and product development process. This
points attention to a criticism that Kenyan biotechnology initiatives
have met previously, deeming them to be exogenous, driven largely by
international private sector interests and supported by the donor community or international private foundations. This situation does to a
large extent mean that local public institutions in Kenya are confined
mainly into field-testing of developed products (see Kirea et al. 2003
who arrive upon the same conclusion).

On the question of whether international collaborations are well-
absorbed into local innovation system for agricultural biotechnology
our survey found in the negative. Primarily, we find that there is a
strong relationship between national strategy for biotechnology development, availability of relevant human skills in the local research institutes, and international collaborative efforts. In other words, if there
were more relevant human skills that could be deployed and if national
and organizational strategies for biotechnology were more clearly set
out and implemented to make capacity-building a priority in international research collaborations (both of which rely on policy capacity, see
section 7.5), strategic involvement of local researchers in international
collaborations could have been effected, which is now not the case.

Our survey also found that most researchers who take part in such capacity building and training are disgruntled by the low state of innovation capacity in the local system which constantly acts as a hindrance to applying their acquired skills to research and innovation activities to the local context.

6.4.3   Strengthening entrepreneurship by providing incentives for collaboration and support

Despite being in compliance with the TRIPS Agreement, Kenya has a very low score of local patent applications when compared to foreign ones, which once again acts as a measure of the low level of local entrepreneurial activity. Patent registrations amounted to 61 in the year 2001, with two of these being registered by residents and the rest by non-residents. In addition to these registrations, the number of international patents in Kenya amounted to a total of 89,180 according to 2002 data available from the WIPO (WIPO, 2007). As our work on cut flowers shows (see section 7.6), intellectual property rights, as provided in Kenya, could have very important negative implications for how local capacity can be built, and Kenya can move higher up from being a mere producer of flowers to a cultivator and creator of newer plant varieties. Furthermore, despite the extremely stringent intellectual property regime, all actors advocating and negotiating for royalty-free access to biotechnologies in Kenya such as ISAAA and the African Agricultural Technology Foundation (AATF) are non-governmental in nature, underscoring the lack of awareness of the impact of the present intellectual property regime.

To enhance access and uptake, there is a need for state created agencies that facilitate technology transfer, and negotiate access to proprietary technologies. The case of Genetic Technologies International Laboratories (GTIL) discussed in box (see Box 6.1) below helps to illustrate some of the issues involved in local entrepreneurship initiatives. Simply put, although GTIL is a very successful private enterprise, its success has been achieved without significant support in several critical areas, including access to technologies and collaboration venues. Providing policy and institutional support for local entrepreneurs through incentives that structure more systematic collaboration between various actors in the biotechnology system, as well as enable development of marketable products needs to be a primary goal of policy reform in this area.

6.4.4   Improve organizational roles and coherence

The survey shows that while most organizations and actors necessary in the biotechnology sector are in place, for example, in the research,

Box 6.1   Genetic Technologies International Laboratories (GTIL) Limited

Genetics Technologies International Limited is a privately owned Kenyan company that started its operations in 1995 and specializes in rapid micro-propagation of healthy superior planting materials through tissue culture (TC) technology. The company has a production capacity of over 20 million plantlets per year, producing a wide range of planting materials including:

Industrial crops such as pyrethrum, sugarcane, sisal, vanilla, coffee, among others.

Horticultural crops such as citrus fruits, mangoes, avocados, passion fruits, pawpaw, and macadamia. Among *flowers*, lilies, eustoma, zantendeseschia, glandiolas and statice are among the varieties that GTIL have engaged in. Most often, GTIL does not produce the flowers varieties but the clientele bring the selected flowers and GITL multiplies/reproduces the number of flowers as required by the client requires.

Food crops include bananas, (i.e. dessert, cooking) and plantains, Irish potatoes and pineapples.

Trees for fuel, wood, furniture, telephone/electric poles. The tree species include eucalyptus, grevillea, acacia, pinus, markhamia, *Croton megalocarpus*, prunus, neem, melia, terminalia, jatropha, moringa, cypress, *Warbugia ugandesis*, and teak

Medicinal plants and herbs such as Artemesia, aloe vera, *Mondia whytei*, bamboo, turmeric, geranium among others.

GTIL came into being as a response to the declining rate of agricultural productivity in Africa and most developing countries, most notably in all crops – food and cash crops. Most surveys carried out attribute this decline to lack of clean planting material to farmers, since many farmers use planting materials infested with disease and pests. There is also lack of improved germ-plasm which could give better yields and resist pests and diseases. Besides, there are other issues such as a lack of farm inputs; high costs of inputs, and the lack of proper technological approach to the farming programmes. Tissue culture seeks to address the inadequacy of clean, disease-free planting materials through a cleaning and multiplication exercise. The enterprise is medium-sized, with 50 employees in total and the staff at the laboratory are trained and supervised by GTIL.

GTIL's Collaborations and linkages

Collaborators/ partners	Nature of collaborations	Remarks
International Service for the Acquisition of Agri-biotech Applications (ISAAA)	For sourcing technologies from across the world	ISAAA sources for technologies but requires organizations such as GTIL to ensure the technologies are multiplied and reach the end-users
The Tree Biotechnology Project	For multiplication of eucalyptus clones	This collaboration was initiated by ISAAA to enable Mondi Forests International import improved varieties of trees into Kenya. The TBP lacked equipped laboratories to handle the cloning
Kenya Plant Health Inspectorate Service (KEPHIS)	For inspection of all GTIL nurseries to ensure materials are disease-free	KEPHIS is a regulatory body whose role is to ensure sanitary and phytosanitary matters are adhered to
Kenya Forestry Research Institute (KEFRI)	For forestry research	KEFRI handled all the technical aspects of the tree biotechnology project
KARI/Ministry of Agriculture (MOA)	For other research and extension services	KARI conducts research especially on food crops while the MOA provides extension services
Horticultural crops Development Authority (HCDA)	For registration of nurseries and inspection	HCDA supervises and oversees the nurseries operations.
African Biotechnology Stakeholders Forum (ABSF) and Africa Harvest Biotechnology Foundation International (AHBFI)	For information dissemination and training of stakeholders in new varieties' acquisition and use	The exact nature of collaboration with these institutions is unclear but both are NGOs involved in biotechnology.
International Network for the Improvement of Bananas and Plantains (INIBAP) – Belgium	For clean planting materials	INIBAP collects and maintains different varieties of bananas and plantains germplasm.

(Continued)

(Continued)

Collaborators/ partners	Nature of collaborations	Remarks
Biotechnology Trust Africa (BTA)	For marketing GTIL's products in western Kenya (Bungoma)	BTA and GTIL have an MoU in accordance with which BTA buys GTIL's products for their (BTA) nurseries in western Kenya (Bungoma) where farmers can easily access the planting materials.

Source: Authors' field survey (2007).

GTIL is fairly independent and does its own marketing, distribution, and financing without any collaboration. Where collaborations exist, GTIL has been the main initiator of these collaborations, and GTIL has faced numerous problems in structuring these collaborations. Except Biotechnology Trust Africa (BTA) with which GTIL has a MoU, the other collaborations are structured rather loosely and are non-contractual in nature. The international linkages provide GTIL with current information in the biotechnology industry, as do most of other GTIL's collaborators and hence the intensity of these collaborations. GTIL occasionally invites specialities from international countries to come and share knowledge and expertise, for example from India and South Africa as a way of improving and marketing GTIL. GTIL admits that more support from the Kenyan government and its designated agencies in both identifying collaborators, advice on structuring collaborations would be very useful. GTIL only produces plantlets and finds it difficult to move beyond to conduct research due to problems of expanding into research without adequate support from other organizations and institutions within the system, and intellectual property issues.

Source: Authors' survey, 2006.

demand, infrastructure, demand, and entrepreneurs domains, it is the organizational competence which is missing. In other words, there is a lack of relevant human skills to steer the organizations into their respective mandates, and to enable them to coordinate their work well in this area. Apart from the latecomer malady of duplicating research efforts across all PRIs despite the limited resources available, as we saw in the case of Nigeria, there have been other instances of explicit waste of research results due to a lack of coordination. The case of tissue culture bananas is one such example, where several such anomalies were clearly evident. Wafula (2000) note that both tissue culture technology and germplasm were imported from South Africa despite the fact that tissue culture work on bananas had been going on in the country for at least seven years. The tissue culture banana case, where the network succeeded in propagating, producing, and disseminating the crop as the cash-crop alternative for those farmers who were not earning well with other cash crops, such as coffee also shows the lack of planning in governmental strategy as to the livelihood impact of this development (See also Harsh and Smith, 2007). This yet again, shows the limitations in biotechnology development activities in the country – which seem to be largely driven by external interests and missing in focus on local priorities and concerns.

More over, most of the actors interviewed observed that the level of involvement of the private sector in biotechnology is still very low and urged that the private sector should be encouraged by enacting appropriate policies and incentives. Some actors also complained of a weak extension system at the grass roots levels and suggested the need to facilitate better flow of information to farmers by strengthening the extension system.

6.5   The role of the state in promoting agricultural biotechnology

Although Kenya has begun to put in place a regulatory framework for agricultural biotechnology and GM crops, the patchwork of laws remains mostly unenforced partly due to Kenya's lack of ability to put into place mechanisms for the monitoring and enforcement as required. There is no strategic policy vision in place to promote biotechnology-led development, especially one that takes into account the technological requirements of the process. While Kenya has a biosafety bill, biosafety committees, and rules and regulations relating to intellectual property, there is no broader vision that links these to science, technology and innovation policy for the sector (or national science, technology and innovation policy for that matter), local needs of farmers, food security and competitiveness. Kenya has a draft Science, Technology, and Innovation Bill that has been discussed since 2006, but this has not yet come into force mainly due to political delays and its Science, Technology, and Innovation Strategy Plan 2008–12 was unveiled only last year.

In this section, we present the regulatory framework as it presently stands for the governance of agricultural biotechnology and GM crops as well as for intellectual property rights on biotechnology products and plant varieties. Despite the developments, the same questions that were asked in the section on interactive learning remain with respect to the learning and innovation aspects and the relevance of international collaborations in building capacity for agricultural biotechnology in a systematic way. As things stand, even if some capacity is eventually built in the sector, it would mostly be a result of several factors coincidentally acting in tandem within the system of innovation, rather than an outcome of vision and policy action of the Kenyan state. A second aim of the regulatory framework would normally be to provide a guarantee regarding the safety and efficacy of the plant varieties' being planted on Kenyan soil. On this point too, the regulatory framework on agricultural biotechnology is at best informal in nature (Clark et al., 2005).

6.5.1 Kenya's institutional and regulatory framework on agricultural biotechnology

The legal framework for scientific and technological research and development are guided by the *Science and Technology Act* (Cap 250) Laws of Kenya. There has been a new Science, Technology, and Innovation Bill under consideration in recent years (2006 onwards). The Act establishes the machinery to avail to the government advice upon all matters relating to the scientific and technological activities and research necessary for proper development and to coordinate research and experimental development. It creates the National Council for Science and Technology (NCST) comprising all the Permanent Secretaries (PSs) of the relevant (scheduled) ministries and 12 other members representing eminent scientists derived from the various scheduled disciplines. However, the Act has no specific provisions on biotechnology and biosafety. This shortcoming necessitated the formulation of a specific framework to address issues of biotechnology development in the country.

The history of development of the national biotechnology policy and legal framework can be traced back to 1990 when the government appointed a National Advisory Committee on Biotechnology Advances and their Applications (NACBAA), which was given the mandate to identify the national priorities on the basis of comparative advantage and the ability to implement traditional methods in agriculture, facilitate access to new germplasm, reduce high costs of agricultural inputs, and promote cheaper access to environmentally friendly alternatives.

Based on this, one of the chief recommendations of the NACBAA some years later, was the need for immediate applications of tissue culture for mass propagation and disease elimination, development of disease diagnostic kits, and the use of biological nitrogen fixation (BNF) (Odame, 2003).

The second major phase in biotechnology policymaking process began in 1993 as part of the Biotechnology Programme of the Netherlands Directorate-General for International Cooperation (DGIS) that sought to develop biotechnology for poverty reduction in Kenya. The programme was split into two major parts: developing specific technologies and enhancing capacity for national regulatory and biosafety capacity. For the former, DGIS set national priorities similar to NACBAA including tissue culture and other low-end biotechnologies noting the need to commence more intensive biotechnologies.

The United Nations Environment Programme – Global Environmental Facility (UNEP-GEF) project facilitated the third phase of biotechnology policy development from 1997 onwards. The UNEP-GEF project, coordinated by the National Council for Science and Technology (NCST) aimed at helping developing countries to develop their national biosafety frameworks (NBF). The UNEP-GEF project was conceived under the auspices of the Convention on Biological Diversity, 1993, to promote the harmonization of biosafety instruments at subregional, regional, and global levels. The two main components of the project were to facilitate the development of national biosafety frameworks in more than 100 countries and assist in the implementation of the frameworks, Kenya being one of them. The first phase of the project that began in Kenya in 1997 included a survey to identify existing applications of modern biotechnology in the country, the extent and impact of release of GMOs, risk assessment and risk management systems and review all existing legislations relevant to biosafety (Thitai et al., 1999).

Until about 1998, biotechnology and related research activities had been governed by the *Science and Technology Act of 1980*. However, the regulations stipulated under this Act were only geared towards trials and were not applicable to field release and commercialization of GMOs. Therefore, the National Council for Science and Technology (NCST) which is the Kenyan government's appointed authority to oversee the coordination and implementation of the biosafety regulations in Kenya, convened a multidisciplinary committee (including the then Permanent Secretary) to develop regulations and guidelines for the country's biosafety system. The regulations which covered areas of GM research and development (R&D), use of all aspects of recombinant DNA technologies, and the release of plants and animals derived through such techniques had the following broad objectives:

- Promote opportunities for the application and exploitation of products of biotechnology;
- Ensure public and environmental safety particularly in accident prevention, containment and waste disposal when GMOs are used in R&D or industrial processes;
- Determine the measure of risk assessment, management and monitoring of operations involving rDNA technologies and products arising thereof.

The adoption of the national regulations and guidelines for biosafety in 1998 provided for the establishment of the National Biosafety System (NBS) and the procedures to follow in setting up such a system. It spells out the NCST as the government-designated body to oversee the coordination and implementation of biosafety regulations and guidelines. The National Council for Science and Technology operates under the Ministry of Education, Science and Technology (MoEST) and was established by the Science and Technology Act (Cap 250) of 1980. The current provisions of the biosafety regulations and guidelines mandates the National Council for Science and Technology to establish the National Biosafety Committee (NBC) whose membership should be drawn from across different agencies. The NBC is charged with the task of drawing policies and procedures besides vetting research applications to ensure compliance with the laid down regulations. The NBC also coordinates and oversees the establishment of Institutional Biosafety Committees (IBCs) in those R&D institutions applying modern biotechnology in their activities. As such the institutional framework for governing GM products and research in Kenya comprises the NCST, the NBC, and IBCs.

The members of the NBC are drawn from across different agencies including government regulatory agencies, scientists, Ministry representatives, research institutes, universities, non-governmental organizations and the Council. Article 6 of the *Science and Technology Act* permits the NCST to appoint and incorporate other committees and states in part that 'the Council may from time to time appoint such working or other committees as it may think fit, and may provide for the regulation of the proceedings of such committees'. Article 6(2) provides for the composition of such a committee to include a member of the Council, who becomes the chair of such a committee and other members of the Council if found appropriate and the council may co-opt any person(s) as an additional member of the committee. The co-opted member doesn't necessarily have to belong to the Council. Scientific and technical expertise for the NBC stems from the scientists representing academia, research institutes, and some government departments. The NBC has powers to appoint task forces, co-opt individuals with the necessary expertise or seek external expert opinions regarding very specific issues where such expertise is lacking within the local population.

Other than the NBC, there are Institutional Biosafety Committees (IBCs) whose constitution allows both in-house scientists and external experts with a mandate to carry out in-house technical reviews and approval of biotechnology research and GMO release applications before they are submitted to the NBC. Kenya's leading research institutions such as KARI and the International Center for Insect Physiology and Ecology (ICIPE) have established their IBCs. KARI, which is the leading applicant for GM research approvals in the country, has an eight-member IBC comprising representatives from KEPHIS, Department of Veterinary Services (DVS), International Livestock Research Institute (ILRI), the University of Nairobi and KARI's in-house members.

Kenya's Biotechnology Policy and Biosafety Bill were drafted and sent to the Attorney General's office in 2004 and were awaiting approval by parliament at the time this survey was conducted in 2006–7. The Biosafety Bill (2005) proposed the establishment of a National Biosafety Authority whose functions shall be to:

- Receive, respond to and make decisions on applications on GM products;
- Establish administrative mechanisms to ensure the appropriate handling and storage of documents and data in connection with the processing of applications;
- Establish a database for the purpose of facilitating collection and dissemination of information relevant to biosafety;
- Identify national requirements for manpower development and capacity-building in biosafety;
- Maintain directory of experts in biotechnology and biosafety;
- Advise institutions and persons on mitigation measures to be undertaken in case of accidents;
- Promote awareness and education among the general public in matters relating to biosafety.

The authority, which will be the national focal point (presently, the NCST is the focal point) will be managed by a board chaired by an eminent scientist appointed by the minister. Other members in the board shall comprise experts in biological, environmental and social sciences; the Permanent Secretaries (or their representatives) responsible for Science and Technology and Finance, the Director-General of NEMA, the Managing Directors of KEPHIS and KEBS; the Director of Veterinary Services (DVS), Secretary of NCST, and the Agriculture Secretary. The chairperson and board members hold office for term of three years and are eligible for reappointment for a further three years. The appointments and their names are published in the *Kenya Gazette*. The regulatory matters under the proposed Act are spread across the regulatory bodies as described above.

Table 6.8 shows the timeline for the development of biotechnology and biosafety systems in Kenya.

Table 6.8 Biotechnology and biosafety policy development timeline in Kenya

Year	Key event relating to biotechnology policy development in Kenya
1990	Government appoints the National Committee on Biotechnology Advances and its Applications (NACBAA)
1993	The DGIS-Netherlands programme begins
1995	*Ad hoc* approval and a permit to import a recombinant animal vaccine
1997	UNEP-GEF phase I begins
1998	Guidelines for biotechnology, biosafety published by NCST
1998	National Biosafety Committee (NBC) formed
1999	Environmental Management and Coordination Act (EMCA) is passed and National Environment Management Authority (NEMA) established
2000	Kenya signs the Cartagena Protocol on Biosafety
2002	Seeds and Plant Varieties Act (1972) amended to accommodate biotechnology
2003	UNEP-GEF Phase II begins – to implement the national biosafety framework
2003	Draft Biosafety Bill prepared
2004	Draft Biosafety Bill submitted to the Attorney-General's office

Source: Authors' survey (2006–7).

6.5.2 The regulatory framework for GM crops and technologies in Kenya

The Kenyan regulatory framework for GM crops comprises five regulatory agencies reporting to different ministries and deriving their legal backing from various Acts of Parliament. The regulatory agencies include: the Kenya Bureau of Standards (KEBS), the Kenya Plant Health Inspectorate Service (KEPHIS), the National Environmental Management Authority (NEMA), the Department of Veterinary Services (DVS), and the Public Health Department (PHD).

The Kenya Bureau of Standards (KEBS) operates under the Ministry of Trade and Industry (MoTI), and is responsible for setting standards for weights and measures, purity and identity. KEBS is the national standards body and is established under the *Standards Act* (Cap 496) laws of Kenya. This Act of 1974 seeks to promote and provide for standardization of commodities and a code of practice. The overarching mandate of KEBS is to ensure consumer safety through setting standards for nutritional content, tolerance levels for food toxins (e.g. mycotoxins) and provide facilities for testing and calibration of precision instruments. In terms of standards, the KS 05-40 labelling for pre-packaged foods exists and covers requirements of labelling all food products. This standard was established in line with requirements stipulated under the codex standards for food labelling. However, the standard doesn't cover genetically modified products.

The Kenya Plant Health Inspectorate Service (KEPHIS) was established as a state corporation under the State Corporations Act (Legal Notice No. 350 of 1996) to regulate all matters of plant health and quality control of agricultural products in Kenya. Its mandate includes overseeing the safe introduction of GM plants, products, and micro-organisms into the country. It derives its regulatory authority from various statutes including the *Plant Protection Act* (Cap 324) dealing with importation of plants and plant products, the *Seeds and Plant Varieties Act* (Cap 326) regulating certification and registration of all seed, the *Agricultural Produce (Export) Act* (Cap 319) governing the exportation of plant and plant-related products from Kenya, the *Suppression of Noxious Weeds Act* (Cap 325) addressing the prevention, suppression, and eradication of noxious weeds among other statutes. Even though the legal notice sets out the role of KEPHIS in biotechnology, this mandate is only limited to plants.

KEPHIS falls under the Ministry of Agriculture (MoA) and has jurisdiction over phytosanitary matters and a full regulatory authority to seize, turn away, quarantine and destroy all materials unacceptable to Kenyan standards. It works very closely with KEBS on phytosanitary issues and routinely inspects and regulates all materials at all entry points through a permit system. KEPHIS also inspects and approves all containment facilities (laboratories, greenhouses and quarantine facilities) and only when satisfied that the facilities meet all the requirements, issues an importation permit.

The Department of Public Health (PHD) under the Ministry of Health (MoH) is charged with regulatory responsibility over health and safety aspects of food and feeds and derives its legal authority from the *Public Health Act* (Cap 242) and the *Food, Drugs and Chemical Substances Act* (Cap 254). Its overall duty is to ensure the public is protected from harmful food, drugs, and other chemical substances.

Matters relating to animal health fall under the jurisdiction of the Department of Veterinary Services (DVS) under the Ministry of Livestock and Fisheries Development. The DVS derives partial authority from the *Crop Production and Livestock Act* (Cap 321) governing the control and improvement of crops and livestock, marketing and processing, *the Veterinary Surgeons Act* (Cap 366) and the *Animal Diseases Act* (Cap 364) dealing with control of animal diseases. In 1994, the DVS permitted the importation of a recombinant vaccine – virus-based rinderpest vaccine developed by the United States Department of Agriculture (USDA) and conducted the testing.

The National Environmental Management Authority (NEMA) operates under the Ministry of Environment and Natural Resources (MENR) derives its regulatory authority from the *Environmental Management and Coordination Act (EMCA)* of 1999 and is mandated to coordinate all development activities and ensure all environmental issues are properly and adequately addressed. Section 53 of EMCA empowers NEMA to make regulations on biotechnology matters as they relate to the environment. At present NEMA does not have any direct role in the GM arena since all GM-related work is still at the research stage, being undertaken in contained facilities. When the GM work goes into field cultivation and commercialization, it is envisaged that NEMA will be more involved in environmental risk assessment and mitigation of any harmful effects.

6.5.3 Kenya's intellectual property rights framework

Kenya is a member of the World Trade Organization and is therefore obliged to implement the Agreement on Trade Related Aspects of

Intellectual Property Rights (TRIPS) Agreement of 1995. Despite the fact that Kenya is exempt from complete TRIPS compliance until 2013, it has already enacted a TRIPS-compliant intellectual property regime. The survey shows that the establishment of the Kenya Industrial Property Institute (KIPI) following the enactment of the *Industrial Property Act* (Cap 509) in 1990 has provided the necessary legal framework for intellectual property protection in the country.

The intellectual property rights in Kenya are covered under four Acts of Parliament namely: the Intellectual Property Act (Cap 509), the Trademarks Act (Act 506), the Seeds and Plant Varieties Act (Cap 326), and the Copyrights Act (Cap 150). The creation of the Kenya Industrial Property Office (KIPO) in 1990 (and its transformation into Kenya Industrial Property Institute (KIPI) with greater decision-making power and authority) following the enactment of the Industrial Property Act was a major strengthening act as far as intellectual property protection in Kenya is concerned.

On the question of plant varieties protection, Kenya had already enacted the Seeds and Plant Varieties Act (Cap 326) far back in 1977, providing for the protection of plant breeders rights. The 1977 Act (which was reviewed in 1991) coupled with the 1994 regulations on the same issue, ensure its compliance with the provisions of the UPOV 1978 convention. Kenya is a member party to UPOV 1978 convention since April 2000, as part of which some of the provisions of the parent legislation and the implementing regulations were both revised.

Two important aspects stand out instantly in this context: Kenya's plant variety protection regime does not focus on the needs of local farmers. The Act has no specific provisions addressing the question of farmers' rights. The 2001 Bill contains provisions that prohibit the exchange of seed among farmers, and therefore does not cater to the needs of local farmers' rights. Second, there are no visible impacts of plant variety protection regime on promoting the nascent local private-sector enterprise in the cut- flower sector, which our survey covered extensively and some of these results are presented in the next section.[3] As early as 1999, the ratio of international to national applications in the cut-flower sector was 91% to 9% (See Grain, 1999), and this wide gap still remains.

6.6   Case study: The Kenyan cut flower sector

Globally, the cut-flower sector was worth approximately US$6 billion in 2006, of which Kenya accounted for 6% in total as opposed to the largest contributor, the Netherlands which catered to 54% of the total global demand (Hornberger et al., 2007). In monetary terms, Kenya's exports rose to US$300 million in 2007, and is one of the fastest growing sectors of the economy (Kenya Flower Council, 2007).

Kenya's cut-flower sector emerged in 1970s and picked up in the 1980s when the leading exporters began to plant commercial rose cultivation for exports to European Union (EU) markets. It gradually transformed from being low value and simple open field flower plantations in the 1980s to high value flower farming in green houses by the 1990s. For example, from 2001 to 2005 cut-flower export grew at a compounded annual growth rate (CAGR) of 27% (Hornberger et al., 2007).

The industry employs more than 1250,000 people directly and a further 2 million indirectly through and related auxiliary economic activities thus contributing to employment creation and poverty alleviation efforts. The industry in characterized by a robust private sector comprising largely large-scale growers dealing in flower varieties requiring huge financial inputs grown under greenhouse conditions such as roses, carnations, statice, alstromeria, and veronica, among others, which require high levels of agricultural technical and managerial skills. The local smallholder growers, however, are mainly confined to summer flowers requiring less capital in-put, managerial and technical skills and can very easily grow under open conditions. At present roses are the top varieties exported from Kenya accounting for over 70% of all flowers grown and exported.

The government's strategy in Kenya has been one of limited participation, where a specialized agency, the Horticultural Crops Development Authority (HCDA) was set up as a state corporation under the ministry of agriculture vested with the responsibility to develop, promote, coordinate, and facilitate the horticultural industry in Kenya. Apart from this, an enabling environment was facilitated through its IPR regime, which provided plant breeders rights to new plant varieties and hence promoted foreign plant breeders to set up cultivation farms in the country and promoting quality standards and incentives for exports. Therefore, market triggers were largely relied on for the growth of the sector and direct involvement of the government, however, has been minimal when compared to other agricultural products, such as coffee.

Our survey covered the cut-flower sector extensively as a case study and 49 farms were administered questionnaires apart from conducting a series of interviews with farm owners and other stakeholders. The sector is no doubt performing well and is the largest earner of foreign exchange in the economy. However, the very factors that were initially responsible for attracting foreign investment for the sector and promoting the growth of the sector are now beginning to act as hindrances to local farmers. Small-scale Kenyan farmers find it extremely hard to diversify their export locations, and to compete in the highly competitive value chain that supplies the Dutch auction system. Our interviews show that local infrastructure and agricultural extension services, such as port and aircraft facilities, ease of transportation to the ports and aircrafts apart from the costs of negotiating licences for the plant varieties, all pose serious transaction costs to the local farmers. Local farmers also complain of the lack of policy and organizational support towards producing local cultivars that are based on African horticultural varieties that ranges from the missing research infrastructure, lack of collaboration between existing research and enterprise as well as risk-attenuating mechanisms for investment into such activities. Local growers frequently quoted the intellectual property mechanism, in the form of strictly enforceable plant variety rights, as a major impediment in their own initiatives to produce their seed varieties.

In sum, we find that policy support for the sector was largely meant once again, to promote the production of flowers only, and there has been no focus whatsoever, at enabling the local sector to emerge as a creative entrepreneur of indigenous plant varieties. The system as it presently stands, clearly does not allow the local farmers to engage in technologically intensive activities to move up the global value chains in production and product development initiatives. This is a loss not just to the cut-flower sector, but to local agriculture as a whole, wherein the cut-flower experience if correctly managed could have been replicated to several other products and more knowledge-based initiatives could have been catalysed. The same is true with incorporation of higher environmental standards in local farms, for which technical support from the government is far from sufficient and effective. There is a lesson here to share for other African countries, such as Ethiopia, which have recently embarked upon the task of promoting local cut flower sectors. Policy focus needs to integrate the needs of attracting investment (such as plant variety protection) with the needs of promoting local activity into more knowledge intensive domains. At a more strategic level, the issue should increasingly move away from merely attracting investments, important as it is, to making the environment more conducive to encouraging greater biotechnology activity in the country and to promote the kinds of investment for building innovation capacity. Without this, African agriculture and Kenya horticulture specifically, will remain a periphery player in the core science-based activities and will continue to playing an auxiliary role relative to large foreign-owned firms as long as it neglects it own local needs and capacity.

6.7   Summing up

This chapter has analysed Kenya's experience in developing its agricultural biotechnology systems of innovation. Our main finding is that while Kenya has been successful in promoting several international research and product development collaborations in this area, and while its legal regime has taken note of the need to have both intellectual property and biosafety regimes, there is no systematic evidence of capacity building as a result of these initiatives. We find that this is due to the lack of clear and cogent policy vision for building innovation capacity in the sector, and the policy framework as it presently stands does not balance the biosafety and intellectual property framework with the science, technology and innovation needs of the sector. A range of factors that have been identified stifles interactive learning, and the case study of the cut-flower sector serves to illustrate how more knowledge-intensive activities are stymied by a lack of strategic focus on innovation capacity to meet local needs. Precisely because of these reasons, although Kenya's largest cluster is the Agricultural Products cluster (with 0.26% market share) has experienced a decline in market share (–0.05%) between 1997 and 2004, according to the World Bank.

7
Comparative Analysis of Innovation Capacity in Latecomer Countries

7.1 Introduction

This chapter re-examines the key findings and perspectives arising out of the country case studies analysed within the sectoral systems of innovation framework that we coded in terms of *innovation capacity* in latecomer economies. The four country experiences for biotechnology capacity sketch out the complex landscape of both hope and distress for food security and economic development among latecomers. There are several strands of comparison that scholars of innovation and development could decidedly take up ranging from analysing general hindrances to innovation capacity across latecomers, tracing successes and extrapolating their causes, and exploring potential for duplication across the country and elsewhere within other latecomers.

However, we choose to focus on two important concerns. First, we revert to the concept of innovation capacity as described in Chapters 1 and 2, which is central to our framework for studying sectoral systems of innovation in agricultural biotechnology in latecomers. Innovation capacity as we defined it comprises the ability of firms and organizations to undertake product and process-related, imitative activities, to build competencies (e.g. universities and public research institutes), the ability of institutions to eliminate distortions in information exchange which is at the root of the missing or ill-formed collaborative linkages in latecomers and, finally, the ability of the state to formulate and implement policies. Latecomers and very latecomers exhibit substantial differences in these institutions from countries at the frontier and these differences are better understood analysed around four key factors – namely, knowledge bases for biotechnology, interactive learning, incentives for translating inventive activities, and state policy capacity. The former two determine the ability of firms and organizations to undertake product and process-related imitative activities and to build organizational competencies, whereas the latter two account for the ability of institutions to eliminate information asymmetries and the capacity of the state to formulate relevant generic and sector-specific policies.

In comparative perspective, this chapter employs the detailed empirical data for a systematic understanding of the nature of agricultural biotechnology and the factors that influence innovation capacity in this sector in a new light, and demonstrates how such evidence is critical at a time when debates on biotechnology, food security, and economic development are mostly informed through general perceptions on what the state of the world in latecomers could be, rather than what it really is.

The four countries presented in this book are at very different stages of innovation capacity in agricultural biotechnology, which adds to the richness of the analysis. Whereas Malaysia is well ahead of the other three countries in terms of having achieved the Green Revolution and providing the knowledge bases for a Gene Revolution, Vietnam exhibits its strength in terms of policy vision that draws upon its previous political character. Nigeria and Kenya, on the other hand, are both struggling with pockets of excellence for the Green Revolution itself. Kenya is particularly interesting because of the several on-going international alliances in the country as part of numerous biotechnological programmes of relevance to the country, East Africa, and all latecomers as such. The factors that contribute to these varied innovation capacities are presented in Figure 7.1 below. In the analysis, we now go beyond the country-level findings presented in Chapters 3 to 6 to comparatively analyse how each one of the countries' performs in terms of these factors, and what forms of intervention may be necessary to attenuate issues of capacity formation.

A second contribution of this chapter will be to link micro-level evidence on agricultural biotechnology to broader issues of policy relevance at the national, regional, and international levels, some of which we have already raised in the earlier chapters. Specifically, we are interested in drawing lessons on whether international rules and regulations on biosafety, intellectual property and trade, and issues of perception of donor priorities impact upon the way capacity is built in agricultural biotechnology in latecomers based on the evidence gathered by us in the different latecomer countries, especially in Africa. Finally, we draw out lessons on how scientific and technological change nuanced as innovation capacity can be applied to the resolution of food supply and

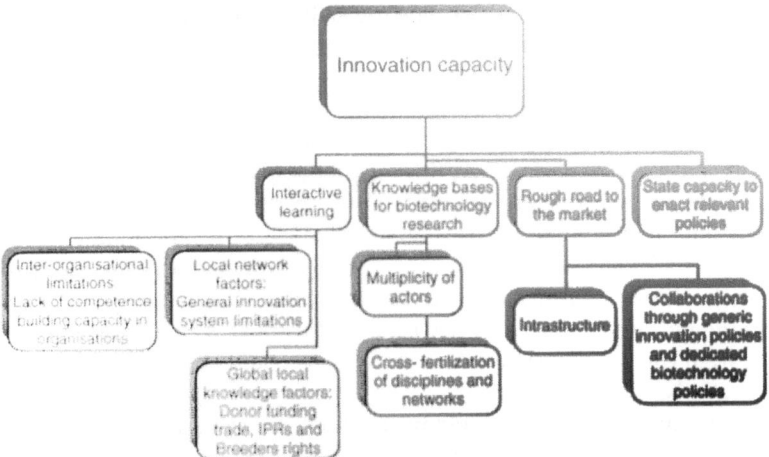

Figure 7.1   Measuring innovation capacity in latecomers
Source: Authors.

poverty particularly in a time of unprecedented global crisis that leaves large swathes of regions in Africa and Asia hungry and malnourished.

7.2   Knowledge bases to promote the Green and Gene Revolutions

In order to sketch the differences underlying knowledge infrastructure in all four countries, we present school enrolment ratios, which are a fundamental measure of knowledge in all the four countries (see Table 7.1). Clearly, the East Asian countries set much greater store on the importance of basic education, and this is reflected in the case of Malaysia which has the highest enrolment rates out of the countries. This is followed by Vietnam, which is laudable especially because of its increasing enrolment rates between 2000 and 2005 for both secondary and tertiary education (up to 16% in 2005 from 9% in 2000). Nigeria and Kenya lag behind their Asian counterparts although the African countries have done equally well in primary but have not been as dynamic in secondary and tertiary enrolment. As expected, the schooling ratio falls from primary to secondary to tertiary in all countries, but the extent of the reduction is least in Malaysia, followed by Vietnam and then by Kenya. The same is true for other conventional knowledge indicators, such as researchers in R&D per million of population and

Table 7.1 Comparative enrolment of countries

Country Year	Malaysia 2003	Vietnam 2005	Nigeria 2004	Kenya 2004
School enrolment, primary (%gross)	93	95	99	111
School enrolment, primary (% net)	93	88	88	76
School enrolment, secondary (% gross)	76	76	35	48
School enrolment, secondary (% net)	76	69	–	40
School enrolment, tertiary (% gross)	32	16	10	3

Source: World Development Indicators (2007).
Note: Data for the most recent available years for countries used.

R&D investments as a percentage of GDP, which although not presented here, have been analysed by us in a broader context.[1]

Other avenues of building knowledge capacity provide useful indicators for comparison. For instance, in the case of foreign direct investment, Malaysia has been extremely successful in attracting foreign direct investment, and effecting related knowledge spillovers in key sectors, such as electronics (see Oyelaran-Oyeyinka and Rasiah, 2009), Nigeria has attracted more natural-resource based FDI and Kenya's own FDI score has not been very satisfactory, with more low-value and high volume-based projects across all sectors (Hornberger et al., 2007).

7.2.1 Reforming university education for agricultural biotechnology

Particularly looking at education of relevance to agriculture biotechnology, the following points are pertinent: while Malaysia has a good university base and traditional strengths in agricultural sciences, with several well-endowed centres of excellence, the same is not true of the other three countries in consideration. Nigeria, Kenya, and Vietnam, for reasons entrenched in their political histories, neglected and even stymied their education systems for long periods. The resultant university education systems are largely outdated. But then, Malaysia has prioritized biotechnology-based education and development relatively late in its country's economic strategies (of the four countries, Malaysia stands out for its prioritization of biotechnology as a lead sector in 2000).

Our data as well as fieldwork interviews reveal that, as opposed to a highly systematic approach taken by Vietnam in rectifying the causes of the education gap for biotechnology, those in Nigeria and Kenya are not systematically aimed at building university disciplines that range across the broad front of what would be important for biotechnology-based research. This lack of systematic focus in guiding the sectoral system of

Table 7.2 Skilled personnel engaged in agricultural biotechnology in Malaysia, Vietnam, and Nigeria

Highest education level	PhD as % of total staff	MSc as % of total staff
Malaysia	39%	44%
Vietnam	16%	46%
Nigeria	51%	31%

Source: Authors' surveys (2006–7).

innovation is critical – our data shows that while Nigeria has the largest number of PhDs (51% of the sample) and Masters' (31% of the sample) in its public research institutes, which is far more than Malaysia and Vietnam for agricultural biotechnology-based work as Table 7.2 shows, this strength is not utilized in national and international programmes on agricultural biotechnology.

Our findings also reveal that the curricula in most faculties of agriculture need reform because they are narrowly focused on scientific/technical aspects of biotechnology and do not cover the full range of legal, regulatory, economic, business, ethical, and social aspects of biotechnology (Eicher, Maredia, and Sithole-Niang, 2006). All of these are essential for capacity development in agricultural biotechnology.

7.2.2 Other institutions for creating new knowledge and using existing knowledge

While Nigeria and Kenya are occupied predominantly in tissue culture (TC) and other low-level biotechnologies, Malaysia and Vietnam are advanced in terms of the biotechnology specializations in both research and product development. Kenya clearly has the basis to engage in more advanced biotechnologies as its experience in international collaborations show, although this remains to be systematically replicated through local initiatives.

Our survey data show that in Vietnam only 12% of all agriculture biotechnology initiatives are focused on tissue culture, while in Malaysia this amounts to 21%.

The point is not necessarily that all latecomer countries need to proceed immediately to invest in higher biotechnologies regardless of their use for local contexts, but rather that:

- The incidence of biotechnology activities at the lower end of the spectrum is a pointer towards the low-technology intensity, whereas sustainable growth of the sector will depend on expertise across the broad range of technologies that encompass agricultural biotechnology.

- The prevalence of TC and other lower-end biotechnologies can still be a positive development if they are being put to use in local agriculture in productive ways, such as the case study of Ghanaian pineapple shows below.

Case study: Ghana pineapple innovation and transformation

In order to illustrate the role of traditional biotechnology techniques we use the innovation and transformation of pineapple in Ghana from a local fruit to an export crop. The case study illustrates the contextual value of the innovation process highlighting the dynamics in the process with various actors becoming critical at various times. In this case study, the National Innovation System (NIS) demonstrates the strengths and weaknesses in the rules of the game, the inter-institutional relationships, the roles and functions of the respective actors.

The transformation of pineapple from a lesser food crop to a major non-traditional export commodity goes back to the 1980s. In the heat of the scarcity of foreign exchange before the Economic Recovery Programme (ERP) of 1983, some Ghanaian entrepreneurs decided to export in order to obtain foreign exchange to finance their businesses. Pineapple was chosen because there was market for it in Europe and Ghana's climactic conditions were favourable for its production. It may therefore be said that the innovation of the crop as an export commodity began with the entrepreneurs with other identifiable stakeholders such as the government, farmers, and scientists, following.

Local scientific research on pineapples began under the National Agricultural Research Project (NARP) with the launching of a Pineapple Project in 1992, aimed at addressing some of the constraints in the production of the crop. The constraints may be categorized as agronomic, biological, and post-harvest handling. The major biological constraints were among others, weed infestation of farms, mealybug, and diseases such as root-and-heart rot and symphilids. NARP enabled some of these constraints to be researched into and improve pineapple production. Currently, annual production is more than triple the 1995 figure with a total of 46,391 metric tons being produced in 2002 earning the country more than US$15.5 million in foreign exchange. The significant increase in the quantity of pineapple produced, which is also reflected in the total hectares under cultivation, is the direct result of the focus on pineapple as an export crop. The volume of export of pineapple is increasing by about 15% annually and it is expected to continue for the next few years with the proviso that existing and emerging challenges are addressed within the framework of the NIS.

Three main varieties of pineapple are exported, namely the Smooth Cayenne, Queen Victoria, and MD2. What has posed a major challenge is the switching from the conventional Smooth Cayenne variety to the MD2. This variety was innovated in Latin America to meet the new demands of the international market. Meeting this challenge is not possible by means of the traditional practice of plucking suckers from mature fruits and planting them. Millions of plantlets have to be produced within a short span of time just to secure the existing market share of the Ghanaian exporters. The technological solution is derived from a tissue culture technique that enables hundreds of plantlets to be produced from a single piece of tissue.

The application of tissue culture technique also meant an evolution for biotechnology in Ghana. Hitherto the technique had been applied in the public research institutions, for example, the Biotechnology and Nuclear Agriculture Research Institute (BNARI) and the Botany Department of University of Ghana on semi-commercial basis. The challenge of supplying MD2 planting materials moved tissue culture into full-scale commercialization in the private sector. Currently, Bomart Farms Tissue Culture Laboratory operates as a private commercial concern employing about 70 people and produces 10,000 plantlets a day. This is several times what was being produced in, for example, BNARI. The institute has plans to step up production to 500,000 plantlets per month. The crucial factor in Bomart's apparent success in meeting the challenge of MD2 planting material is the strong linkage it has with expertise in the Botany Department of the University of Ghana.

The evolution of institutions also occurred in the private sector to meet the national goal of poverty reduction articulated in Ghana's Vision 2020. Farmapine Ghana Limited began operating in September 1999 under a scheme the World Bank introduced known as the Farmer Ownership Model (FOM). It was an innovative concept initiated to facilitate the growth of pineapple exports at the grass roots of society. As a 'bottom-up' approach to a non-traditional export, it was considered to be capable of increasing the incomes of the small-scale rural farmers. Though this approach is yet to be proven, Farmapine is apparently doing well by exporting 4206 metric tons of pineapple in 2000 rising to 6148 metric tons in 2001, and dipping only slightly to 6025 metric tons in 2002. But Farmapine is yet to comprehensively address the MD2 planting material challenge in the manner that Bomart has done – the crucial factor being the absence of an established linkage to tissue culture competence as in the case of Bomart.

Another company with no links to the scientific community in its pineapple operations is Blue Skies. This is a fruit processing company licensed to operate as a Free Trade Zone company and produces fruit salads and juices for the European market. It employs more than 800 people with approximately 40 of them possessing scientific and technical expertise – food scientists, agronomists, and engineers. The large technical manpower seems to have given it self-sufficiency and capacity to innovate particularly on the basis of the institutional knowledge of its markets. Currently Blue Skies exports about 135 different fruit products to various supermarkets in Europe.

The policy and regulatory environment also has its own challenges. One such challenge comes in the form of a demand to meet quality specifications on the international market. For example, exporters have to produce to meet Europe's Good Agricultural Practice (EUREP-GAP) standard specifications. These specifications require that there be a quality assurance process, which starts at the farm level and is carried all the way through the packhouse, processing, and industrial plant levels. In practice this means the various actors must change the way they do things. Farmers with training must keep simple records essential for the inspections. Exporters must impose a strict regime of quality control or risk losing their investment in exports. More importantly, international inspection companies have found themselves with instrumental roles in the export process.

It is important to recognize that not only external influences determine the policy and regulatory environment. In order to influence the policy environment the exporters have formed the Sea-Freight Pineapple Exports of Ghana (SPEG) – a powerful lobby group as SPEG probably accounts for about 90% of pineapple exports from Ghana. It has been able to engage international agencies in programmes for the development of the pineapple export sector. For example, to assist members to meet the EUREP-GAP standards, it has linked with the German Technical Cooperation (GTZ), USAID through AMEX International Inc., the Natural Resources Institute (NRI), and the Plant Protection and Regulatory Services Division of MOFA. Such networking and linkages constitute an important aspect of the functions of a vibrant NIS.

Another major challenge is meeting the overall national development agenda of poverty reduction and wealth creation. The national vision is to attain middle-income status with the specific aim of achieving a per capita income of at least US$1000 by 2010. This vision is also in line with the interests of international development partners such as the World Bank and international donor agencies. Poverty reduction and an emphasis on empowerment of the vulnerable groups of the society – especially women – have become the driving forces for launching some of the projects contributing to the growth of the pineapple sector in Ghana. For example, development partners support out-grower schemes, which give credit facilities to enable farmers to produce for the exporters and provide training for farmers. However, the issue of whether the poor farmers are getting what they should get or that much more is concentrated in the hands of the exporters, are questions for interested parties to answer.

Innovation systems are as effective as the linkages between critical actors as well as within the actor categories. In Ghana's NIS, it appears some of the intra-association linkages are defective. For example although SPEG is making impact in the pineapple sector, most exporters want to solve their own problems thereby weakening intra-group relationships. This attitude is derived from what appears to be a general Ghanaian business culture of being 'lone rangers' and shying away from business partnerships. This further weakens in the innovation system.

In conclusion, the challenge for pineapple farmers and exporters is not only about the multiplication of an existing marketable pineapple variety. There is also the need to go beyond multiplying the MD2 variety and to conduct research into local varieties having the qualities the international market demands. It is a huge innovation burden that requires marketing strategies to condition international consumer tastes to accept the Ghanaian varieties. The innovative capacity to address the challenges must be evolutionary and not *ad hoc*. Any capacity must have its own inherent dynamism and be able to respond in real time to market demands. This is especially so given continuously changing regulatory standards such as EUREP-GAP. The capacity to meet these changing standards no matter how rapidly they occur ensures a competitive advantage over competitors. For such a system to develop, the weaknesses must be resolved, inter- and intra-actor relationships must be strengthened, and any source of dysfunction must be excised.

7.3 Science technology and innovation investment

What tends to separate two latecomer countries more than any other factors that foster development is the intensity and depth of investment in human and non-human resources at firm and farm levels. Agricultural biotechnology relies immensely on the STI investment into agriculture, on the one hand, and then into biotechnology (or the bases for new technologies in general), on the other.

7.3.1 Firm and farm-level constraints in latecomers

Efficiency of agricultural production in latecomer countries is decided by several factors acting in tandem. Evidence on transgenic crops, such as Bt cotton yields in latecomer countries shows that abiotic stresses such as rainfall levels can lead to fluctuating yields. Apart from such extraneous influences, a lack of resources, equity issues, and access to competitive markets are other factors that can affect efficiency of agricultural production.

In *resource-poor farming*, which is defined as farming that uses limited access to production and managerial resources (Spillane, 2002, p. 68), there are several factors that cause efficiency losses in agricultural production. Even when traditional crops are used, the lack of external resources, the lack of organization skills in smaller farms and a lack of fertility of agricultural lands all contribute to less-than-average crop yields. Even where adequate resources may be present, access to resources may not be designed on rules of equity. Where resources may be adequate and access is available, farmers may lack completely the managerial competence required to handle farming in an efficient way. To add to all this, since small-scale farming is mostly done on a subsistence basis, marketing of agricultural produce is not the general measure of performance. This could mean that even where resources may be available and farmers have managerial competence, the lack of access – or poor access – to competitive markets where agricultural produce can be traded may hinder their potential. Clearly, good management at the farm level which is the ability to manage farming with high resource inputs and good managerial skills so that this results in increased profit margins is a precondition to consistently obtain higher yields through the use of both traditional or new biotechnology-based crops. The potential of biotechnology-based crops to produce higher yields in latecomers in the mid- or long term will be critically affected by the presence (or rather the absence) of adequate managerial and production resources at the farm level.

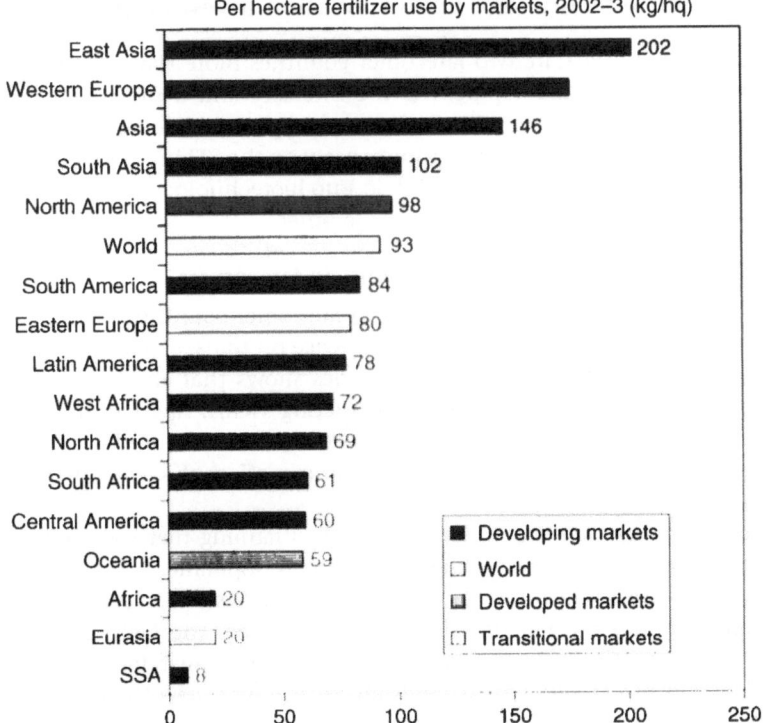

Figure 7.2   Regional per hectare fertilizer use
Source: Derived from FAO data.

Figure 7.2 shows per hectare fertilizer use in different regions of the world as a measure of resource-poor farming practices of the kind evident especially in latecomers in Africa. The figure also helps to frame the differences between Asian and African latecomers in this regard, where the former are far more advanced.

Case study: Oil-palm production in Malaysia and Nigeria

Here, we highlight the contrasting experiences of two of the countries. How, on the one hand, Malaysia sustained investment in human capital and R&D (no matter how imperfectly) and, on the other, how a *laissez faire* attitude on the part of Nigeria has led to a fast-forward decline in a particular sub-sector of agriculture, namely, oil palm production.

Malaysia's combined growth of researchers and expenditures resulted in a doubling of average spending per scientist from US$175,000 in 1981 to US$344,000 in 2002. These averages, however, mask considerable variations among the sample agencies. Spending per researcher at the three prominent commodity boards and government agencies, for example, was above US$500,000 in 2002. In contrast, agricultural research expenditures at, for instance, the Malaysian Agricultural Research Institute (MARDI) measured US$244,000 per researcher the same year, and three of the four Sabah and Sarawak-based public agencies had spending-per-scientist levels below US$200,000. The variations are explained in part by the focus of state research agenda which places priority on key commodities, namely, oil palm, rubber, timber, and cocoa.[2]

In 2002, Malaysia invested US$1.92 on agricultural research for every US$100 of agricultural output. This represented roughly 80% more than the corresponding 1995 ratio of 1.07. The huge increase in Malaysia's agricultural research intensity ratio during 1995–2002 is the result of the deliberate support for public agricultural research, combined with a small decline in the country's AgGDP over this period. The 1995 ratio for Malaysia was higher than the reported 1995 average for Asia (0.63) and the developing world (0.62), but lower than the corresponding ratio for the developed world (2.64) (Pardey and Beintema, 2001).

On the part of Nigeria, government still conducts a large part of agricultural research. This is to be expected and, as with many other African countries, there is little research capacity in the private sector. In addition to government RDIs, a number of agriculture-based universities have emerged as the country expanded university education (agricultural R&D capacity in the higher-education sector grew from 8% in 1971 to 19% in 2000). However, quantitative expansion has not been commensurate with capability growth.

There were approximately 5000 FTE researchers within public agricultural research agencies in five African countries in 2004 with Nigeria having some 1400 of these.[3] However, while Nigeria has the largest number of FTE researchers in Africa (11% of the region's total), its spending at only 7%, is disproportional and relatively small compared to its national wealth. This figure points to Nigeria's poor commitment to STI and also demonstrates the poor environment in which the countries' researchers have to work in comparison with a country such as Malaysia or even South Africa.

Evidently, one critical weakness of the Nigeria STI system is poor funding of innovation activities. Financial allocation to university-based as well as RDI-based work has declined relative to growing national revenues and in relation to other activities. The poor funding of innovation is symptomatic of a broader failure to deploy innovation in areas such as seeds and fertilizer to raise yields, see Figure 7.2 on capita fertilizer consumption: 8 kg/ha for SSA contrasted with 103 kg/ha for South Asia.

Significantly, Nigeria has a higher percentage of researchers at research institutes such as the Nigerian Institute for Oil Palm Research (NIFOR) compared with Malaysia where almost half of the research force (47%) hold PhD degrees. However, this has not translated into any significant effect on the industry in the country when compared with its counterpart institute in Malaysia where the Malaysian Palm Oil Board (MPOB) has generated more than 300 technologies and products. The MPOB has also filed 200 patents – as at 2007 – with over 30% of these having been either sold or commercialized. NIFOR, on the other hand, has generated only five products aside from its well-known variety of planting material and no patents have been filed.

A variety of reasons tend to explain the differential performance of the two countries in the face of the given human resources in both. Clearly the early recognition of the need to promote close relationship between research and industry influenced the direction of state policies that drives the key sectors in Malaysia. There is also the role of historical path-dependence whereby colonial government policies in the oil-palm sector virtually precluded the participation of the private sector.[4] This was the case until early 2000. The government of Malaysia, on the other hand, systematically deployed policy to induce closer actor coordination.[5]

Table 7.3 shows the comparative productivity levels for Malaysia and Nigeria and indicates significant productivity differences between the two countries. From the perspective of latecomer catch-up the gap poses both a challenge and an opportunity for a very latecomer such as Nigeria to use biotechnology to raise its level of productivity. This is shown by the ideal yield conditions achieved in Nigerian research organizations. The table highlights that Nigeria has a considerable distance to go in achieving what Malaysia has already done at the farm level.

Table 7.3 Productivity of the oil palm sector in Malaysia and Nigeria

	Nigeria (farm-level)	Nigeria (ideal in research organizations)	Malaysia (farm-level)
Tonnes bunches/ha	8	25–30	30
Oil yield	12%	18–22%	22%
Tonnes palm oil/ha	2.5	4–7	6.6
Tonnes kernel oil	0	0.2–0.3	0.33

Source: Authors survey (2007); NIFOR (2007), and MPOA.

7.3.2 Reliance on donor funding for science, technology, and innovation: A misplaced emphasis?

Weak governmental research funding leads to extreme reliance on donor-funded research, a point that was repeatedly confirmed by the data in all countries. While Malaysia and Vietnam have biotechnology research programmes that are extremely well-funded by the government, in Nigeria and Kenya the researchers at PRIs and university departments depend primarily on donor-funding to conduct often the most basic experiments. This reliance on donor-funded research is deeply entrenched, leading to researchers and organizations redefining their areas of specialization even – at times – to attract donor funds: the only ways to fund research. This is detrimental to long-term innovation capacity-building, especially in the case of agricultural biotechnology, because our surveys reveal that an exclusive focus of donors in the area is not to build innovation capacity *per se* but rather to build capacity to enact biosafety frameworks and to enable biosafety monitoring in the countries. This would represent a commendable goal, if it came over and above a country's own sustained effort to build innovation capacity in agricultural biotechnology to meet local needs. However, it is far from adequate when it is one of the only biotechnology-based activities of scientific rigour being conducted within a country.

The lack of donor focus on innovation capacity in latecomer countries is an important issue that needs to be addressed and UNCTAD (2007) notes – based on a review of all donor aid programmes between 1998–2000 and 2003–5 in least-developed countries – that aid for science, technology and innovation is a low priority. The annual disbursements for the development of advanced and specific skills and for research during the period 2003–5 constituted US$727.7 million, which was equivalent to only 3.6% of total disbursements (p. 167). Other conclusions of the report relevant to our purpose are: between 1998–2000 and 2002–5, the aid for development of higher skills increased, although this was to a greater extent aimed at higher education and less towards research institutions, aid commitments for research over the periods of time remained pretty much constant although their composition varied. And, most importantly, *aid commitments to agricultural research halved over the period of time in all least developed countries* (pp. 168 and 169, emphasis added).

In a closer scrutiny of the main directions of aid research, Farley (2007) identifies four major kinds of donor support for all latecomer countries:

1. *Global or regional public goods initiatives*: In the case of agricultural biotechnology, this includes projects such the insect-resistant Maize for Africa project (IRMA) with the International Maize and Wheat Improvement Centre (CIMMYT) and the Kenya Agricultural Research Institute (KARI) based on financial support of the Novartis Foundation for Sustainable Development (which later gave these responsibilities to the Syngenta Foundation for Sustainable Agriculture, SFSA).

2. *Initiatives that deepen domestic STI capacity*: These include projects for developing human resources, supporting domestic research institutes, improving universities or supporting the development of technological capabilities at the enterprise level, and are more general and deal with the local network factors that improve interactive learning on the whole.

3. *International linkage initiatives*: The emphasis of donor-funded activities, projects, and programmes in this cluster is on the creation of capacity to link up with global and regional knowledge networks. The biosafety programmes fall into this category.

4. *Integrated initiatives* that seek to strengthen innovation systems or to integrate the multiple dimensions of STI capacity-building addressed in the above three forms of donor-aid.

The reasons for low prioritization of donor aid on STI capacity is rather a result of several donor-side priorities, including national Poverty Reduction Strategy Papers (PRSPs), as the report has identified. Latecomer countries, especially those in Africa, tend to have low bargaining power (which is exacerbated by their lack of information and policy competence) on strategic directions for sectors such as agricultural biotechnology, which further compounds the issue. Furthermore, our surveys also reveal that donor programmes and international collaborations often tend to be poorly linked to the local system, as in the case of Nigeria and Kenya, whereas Vietnam demonstrates substantial policy competence and planning in harnessing international linkages.[6]

7.4   Interactive learning in agricultural biotechnology: What matters and what not?

We identified three specific sets of issue that limit interactive capacity among latecomers that we sought to capture through our empirical work. The first of these is that *inter-organizational limitations* could stem from the weak institutional basis that organizations tend to have in latecomers – mostly, as a result of organizations that have no competence building power within themselves because they are established without mandates, or without adequate funding, or have overlapping structures, or have no substantial human capacity to draw upon in order to function in an effective and outcome-oriented way. Second is what we called local network factors that tend to reflect in the resource-constrained environment for innovation. There is a strong symbiotic relationship between the national innovation system and the sectoral system – a large number of the institutions and incentive instruments that promote interactive learning span across sectors and are generic, and when these are not well-structured, sector-specific institutions may work at cross-purposes with the national system. Whereas fast followers have more skilled professionals, a better public research infrastructure, a good enterprise sector, relatively sound policy competence, functioning risk instruments and collaboration incentives, latecomers and very latecomers suffer from the lack of these very basic framework conditions. Finally, the *global–local knowledge factors*, that is, the external impetus to local biotechnological activities that shape the perception of activities related to knowledge creation for instance, intellectual property rights and breeders' rights, sanitary and phyto-sanitary measures, biosafety requirements, and public perceptions of GM technologies.

7.4.1   Inter-organizational limitations in agricultural biotechnology systems

A lack of manpower impacts upon organizational competence in profound ways; the ability to conduct relevant research being only one of them. One of the features of effective innovation systems is the way organizations beyond the state are playing their role in the creation and development of research and entrepreneurial opportunities. In addition, role flexibility is also important, as highly compartmentalized and rigidly defined roles do not allow organizations to reconfigure and respond flexibly to changing circumstances. For example: Is the public sector concentrating too much on technology development and not enough on its role in providing supporting structures for innovation such as credit and training? Is the private sector being too risk averse in its assessment of new technologies and how these could be utilized in product development initiatives? These aspects call for constant monitoring and re-evaluation of organizational mandates and thus flexibility.

Our surveys and interviews suggest that, although all countries have embarked upon the creation of sector-specific institutions, these are not very well coordinated in Nigeria and Kenya and that the organizations are ineffective in performing their functions largely because of overlapping and incoherent mandates. Most new organizations set up for performing biotechnology-related functions have no clear mandates, no funding, and have to deal with the ambiguities of operating within an uncertain policy framework. This exacerbates the inefficiency of the organizational structures that are already over-burdened by staff limitations and low actor morale. The institutional inertia thus becomes internalized and is an issue of equal importance for both new organizations as well as those that have existed for several decades in the context of late development.

Rectifying this, and providing a more robust basis for interactive learning through organizational collaborations in the sectoral system for agricultural biotechnology, will require policies that not only delimit a set of objectives (as in the National Biotechnology Policy and the Biosafety Framework) but also target the elimination of basic hurdles to innovation. These include:

(a) *Physical infrastructure and networking*: Autonomous telecommunications faculty such as telephone, fax, and Internet in the universities falls largely below what may be required for good research and teaching work, getting to know and linking up with similar programmes with other universities and institutions within the country or the rest of the world and for dissemination of research efforts. There is critical shortage of power and water; a situation that calls into question the commitment of governments to research investment.

(b) *Knowledge infrastructure*: High quality research requires equipment and investment in sector specific facilities and laboratories at the national and regional levels. Mostly research equipment is not available and even laboratories created to facilitate them depend on the availability of foreign grants for their survival. Basic chemicals and reagents are hard to find and graduate students often have to rely on own funding to carry out experiments. Researchers and their initiatives get stunted by lack of funding and the resulting intellectual isolation becomes internalized into informal norms and codes of conduct.

(c) *General and specific innovation incentives*: Sectoral policies are important, coordination between sectoral policies, especially those that link research in traditional sectors and new technologies are critical.

The comparison between the two countries shows failures (albeit of different kinds) on linking their investments into biotech R&D (in the public sector) with the enterprise sector. There are no specific incentives created in the local innovation systems that promote commercializing of research results. Newer and more dynamic policies and incentives are required to bridge the gap between research and production (private-sector) activities in these countries.

The relevance of these factors cannot be stressed enough, and this is what separates the two Asian countries from the two African ones in the comparative analysis in terms of interactive learning and collaborations. Other results of significance are:

(a) Human capital affects positively and significantly both foreign and local collaboration. In other words, the more the presence of staff in local organizations with MSc and PhD degrees, the greater the incidences of collaboration.
(b) Across all countries, collaborators always perform better than non-collaborators in terms of product and process development;
(c) Malaysia and Vietnam provide a network of much better structured initiatives than Nigeria and Kenya. In Nigeria and Kenya, foreign collaborations tended to be better structured, evenly funded through all phases of the project in question, and there were clear demarcation of roles of responsibilities of all partners. In contrast, local collaborations tended more to be affected by the distortions in the local innovation systems, especially those related to funding and insufficient incentives in the policy framework to structure mutually beneficial collaborations.

We find extremely diverse and different explanations for the low innovation performance of the two countries we studied in Africa, and this points to the challenges of devising a uniform set of policies in latecomer contexts or *to classify issues of African innovation into a generic set*. While both Nigeria and Kenya have failed in different ways to promote sectoral systems of agricultural biotechnology, the causes of failure vary. Nigeria, as the previous section shows, has a very high rate of researchers with PhDs and Masters degrees in the universities and research institutes, however, no coordination of research with industrial product development activities is evident and also little effort is being made to strategically attract international partners. In Kenya, evidently the shortage of human capital stunts the growth of the sector in significant ways and poses hindrances to participate and build capacity through on-going international collaborations. The comparison between these two countries and the rest of the countries in the region is also revealing, For example, the tables below show that Kenya has more organizations engaged in biotechnology than its other East African counterparts, although the country underperforms relative to the attention donors and other actors devote to it.

7.4.2   Local network factors

One of the most significant results of the book is that performance in agricultural biotechnology remains dependent on the state of both the national and agricultural innovation systems in a country simply because it builds further upon these existing strengths. The three agricultural biotechnology pre-requisites: knowledge capital for biotechnology-based research, networking capabilities to form hybrid organizational structures, and evolved regulatory and policy capacity – all rely on the existence of institutional structures for

Table 7.4   Biotechnology infrastructure in selected Eastern African countries (2003)

Technology	Ethiopia	Kenya	Tanzania	Uganda	Total
Tissue culture	1	17	5	6	29
Molecular markers	3	2	0	5	10
Recombinant technologies	1	5	2	0	8
Transformation	0	2	0	0	2
Biofertilizers	1	2	0	1	4
Biopesticides	1	1	0	1	3
Fermentation	0	3	0	0	3
Total	7	32	7	13	56

Source: Wekundah (2003), quoted in Bananuka (2006).

Table 7.5   Biotechnology human resource capacity status in selected East African countries (2003)

Technology	Ethiopia	Tanzania	Kenya	Uganda	Total
Tissue culture	11	17	28	10	66
Molecular markers	3	0	2	13	18
Recombinant technologies	1	21	14	5	41
Transformation	0	2	2	0	4
Total	15	40	46	28	129

Source: Wekundah (2003), quoted in Bananuka (2006).

agricultural innovation that enabled the Green Revolution. This once again points to the path-dependence of innovation capacity; a point repeatedly made by the innovation studies literature.

The comparative analysis again particularly helps to highlight the importance of the underlying infrastructure for agriculture to perform biotechnology-based activities. The agricultural support systems are precisely the strengths in both Vietnam and Malaysia, however, the basis for the Green Revolution has been in place and that undergirds the emerging dynamism of the biotechnology system of innovation. We find that Malaysia – a relatively late entrant into biotechnology – is still managing to cope better than Nigeria or Kenya, mainly because of the path-dependency of these investments. The same cannot be said of Kenya and Nigeria. In Kenya, for example, the performance of the agricultural sector – like that of the rest of the economy – remains below potential as a result of a number of factors. The sector has suffered from drought since 2000 and the terms of agricultural trade have deteriorated with increases in prices of agricultural inputs and decreases in prices for agricultural produce. The government initiated the Strategy for Revitalisation of Agriculture (SRA) in 2004 and introduced a sector-wide Kenya Agricultural Productivity Project (KAPP) to support essential activities such as research and extension services to farmers, but has not been of much help. In Nigeria, once again, the share of agriculture in both aggregate GDP and non-oil GDP increased only marginally during the period 1991 to 2000 (Manyong et al., 2003) and the same trend continues. There are extensive yield gaps in major crops planted in the country and both the agricultural policy and the agricultural extension policy need a re-look in the context of local food security.

For instance, the differences in yield level of major crops between Nigeria and Malaysia such as oil palm is considerable and although there has been some limit on growth, progress has been made as a result of innovation in planting materials, which is around 20 tonnes fresh fruit bunch (FFB) ha/yr. According to scientists interviewed in course of our work, it was expected that up to 35% or even higher yield of oil palm is realizable if clones are used; which will raise FFB yield and improve the oil extraction rate (OER) to 35.25 (FFB yield of 35 t/ha/yr and 25% OER) respectively. In contrast, FFB in comparative large oil palm estates in Nigeria vary from 10 to 14, about 50% that of Malaysia.

Other limitations of the national innovation system, especially those related to physical and knowledge infrastructure and collaboration, all impede the emergence of biotechnology-based capacity, despite several sector-specific initiatives in these countries.

7.4.3   Global local knowledge factors

All of the global factors – trade, intellectual property, breeders' rights, and biosafety – have the potential to impact agricultural biotechnology in both positive and negative ways depending on the latecomer country's relative stage of development. A vast literature has examined this issue (Tansey, 1999; 2000, among others). The ways in which latecomers can integrate them in mutually beneficial ways into local systems of innovation is often inhibited by their own policy capacity that does not allow for a clear assessment of all the benefits and risks associated with particular policy choices.

We begin here with international linkages, as opposed to donor programmes, and the experiences of the four countries under consideration. While Nigeria has not focused much on attracting international collaborations in this area, Kenya has been successful in engaging several international partners, although on an *ad hoc* basis. We use the term *ad hoc* to describe Kenya's experience because our survey finds little evidence of a systematic capacity-building impact of these initiatives on the biotechnology system of innovation for agriculture within the country. As we noted in the Kenya chapter, this is due to the lack of strategic state vision on how to harness these collaborations towards local capacity and local needs, and the lack of scientific manpower. Vietnam, on the other hand, makes for very interesting comparison wherein the otherwise sporadic international collaborations are being systematically integrated into the national agenda for agricultural biotechnology through extremely simple initiatives. The government of Vietnam offers scholarships to promising scholars to undertake higher studies in frontier countries, and these government-funded graduates (and PhD students) return to take positions of importance in local research institutes and carry forward the mandate of maintaining serious academic exchange and research collaborations with foreign institutions where they studied, including Germany, France, and the UK. The Vietnamese government has also put in place several additional incentives for researcher exchange, joint projects, and strategic collaborations, which all add to the enthusiasm of the local researchers who return to the country to pursue biotechnology-based research. Our survey found evidence of this in each and every public research institute and local firm we interviewed. This is a far cry from the disillusioned research environment in which individual researchers in Nigeria and Kenya are wasting their potential and struggling to make research projects come through, and have neither the means nor the dedicated students to sustain them over time.

Intellectual property rights may or may not be able to create the appropriate incentives for innovative activity in latecomer countries. There is enough evidence in other sectors that intellectual property rights *per se*, do not really matter for innovation in latecomers simply because the technological capacity of local actors is much below the threshold where such an incentive would begin to become pertinent (see Gehl Sampath, 2007; Gehl Sampath, 2009).

Specifically in the context of plant variety protection, as of 2009, 20 of the 67 members of the UPOV Convention are developing countries.[7] This does not include other countries such as Bangladesh, India, and Thailand who have adopted modified versions of the UPOV conventions (Musungu, 2007). Eaton et al. (2006) compare the impact of plant variety protection regimes on five countries (China, Columbia, India, Kenya, and Uganda) and conclude that there is no clear evidence garnering to suggest that plant variety protection is bearing positive results on local innovation activity. The recently released report of the International Assessment of Agriculture Knowledge, Science and Technology for Development (2008) notes that latecomers hold a mere 3% of all patents worldwide, 80% of these are in the hands of corporations and individuals in the highly industrialized countries and the world's top five biotechnology firms control over 95% of gene transfer patents.[8] While this might sound like an intuitive result, it is important to focus on another result of the study – the importance of plant variety protection is likely to grow in the future for both agricultural security and innovative activity, and the need to balance these with ongoing developments.

A point of note, while considering intellectual property rights, is that the main debates in Nigeria and Kenya on biosafety and plant variety protection appear to be far removed from their own science, technology and innovation strategies and how plant variety protection and biosafety are a means to that end. In Malaysia and Vietnam, there is a significant effort to create this precarious balance. Our surveys and interviews lead us to conclude that a large factor that influences this is organizational competence – many of the African policymakers often do not understand what is at stake when speaking of biosafety plant variety protection.

7.5 Rough road to the market: The limitations of infrastructure and dedicated policy instruments

Drawing further on the three factors that make it difficult for countries to move from research into product development identified in Chapter 2 (a lack of physical and scientific infrastructure, a lack of interlinkages, and a lack of dedicated biotechnology policies), we make a number of observations. A most critical point is physical and knowledge infrastructure, which plays a key role in the way actors harness the capacity of the system in moving products along to the commercialization stages. For instance, Malaysia generates more than 100,000 kilowatts of energy for 26 million people whereas Nigeria generates less than 6000 kilowatts for a population of 140 million, a situation that constrains manufacturing and export capacity as industrial plants and farm capacities are rendered idle much of the day – with power outages in the latter. Consider for instance Table 7.6 below which presents a bivariate estimate of the key factors that affect the ability of organizations to commercial inventive activities.

The evidence suggests that human capital plays a positive and significant role in product innovation, but this impact is limited to the laboratory and therefore fails to affect process innovation. Local capacity development (through training) positively and significantly affects both types of innovation, while foreign training is related positively and significantly to process innovation. From our earlier analysis on the sub-optimal use of manpower and poor infrastructure, skilled manpower, and the state of water supply (same as power supply) affect negatively and significantly the propensity for product innovation, while technical collaboration is ranked by all respondents to plays a positive and significant role in product innovation. In sum, available manpower and state of water supply display inverse relationship with innovation performance, in other words, instead of supporting process innovation, they tend to hinder product innovation (negative effect).

The same result, with slight variations, accrues when we conducted similar analysis across latecomer countries – policies for finance and additional support (such as IPRs), physical and scientific infrastructure and collaboration variables – are the three critical determinants of product and process innovations (see Gehl Sampath and Oyeyinka, 2009).

In cases where there were successful results, these again were the reasons why the success was limited. For example, in Kenya, despite a thriving cut-flower sector, there is neither a replication of this successful experience in other forms of agriculture activity; nor a move into more dynamic activity beyond simply producing flowers. A major cause of the problem is the inability of the Kenyan system to produce its own horticultural variety (based on African flora or already existing varieties such as roses). While breeders' rights are a limitation in this regard,

Table 7.6 Bivariate probit estimates: Nigeria innovation capacity factors

Variable	Coefficient	(Std. Err.)
Product innovation		
Human capital	0.489**	(0.234)
Foreign training programme	0.008	(0.244)
Local training programme	0.586**	(0.276)
Govt. innovation incentive	0.742***	(0.283)
Skilled manpower	−0.558**	(0.245)
Technical collaboration	0.607**	(0.263)
Laboratory facilities	0.162	(0.283)
IP protection	−0.349	(0.266)
Quality of ICT	−0.214	(0.243)
State of power supply	0.078	(0.436)
State of water supply	0.649*	(0.332)
Other policies	−0.554	(0.794)
Government funding	0.521**	(0.255)
Intercept	−1.547***	(0.304)
Process innovation		
Human capital	0.147	(0.230)
Foreign training prog.	0.480**	(0.242)
Local training prog.	0.527*	(0.274)
Gvt. Innov. Incentive	0.755***	(0.272)
Skilled manpower	−0.079	(0.238)
Technical collaboration	0.269	(0.256)
Laboratory facilities	0.090	(0.274)
IP protection	0.147	(0.250)
Quality of ICT	−0.359	(0.238)
State of power supply	−0.153	(0.408)
State of water supply	0.336	(0.295)
Other Policies	−0.896	(0.728)
Government funding	0.258	(0.253)
Intercept	−1.671**	(0.304)

Source: Authors' field work and analysis.

as the Kenyan chapter in this book has noted, there are other serious impediments to the process. These include:

(a) the inability of the research institutions and university departments to conduct research that suit the needs of the local private sector;
(b) the inability of the local private sector to rely on financing opportunities within the country to branch out into other activities spanning beyond production;
(c) the lack of physical infrastructure for flower growers that they have to surmount in order to get their products to the markets in

Europe – however simple this may sound, in the absence of functioning extension services that growers can rely on, little resources are left to deal with technological upgrading, and

(d) the lack of policy emphasis on moving the cut-flower sector from production to more knowledge-intensive activities.

The most relevant point of comparison was the enterprise sector: as opposed to Nigeria and Kenya that hardly possess a local enterprise sector in biotechnology, both Malaysia and Vietnam have numerous biotechnology firms, well situated to develop products of relevance to their local contexts. Table 7.7 below shows the results of our survey on the question of product development. We asked firms in both Malaysia and Vietnam the source of technology for the products that are developed. As the table shows, in Malaysia, most firms relied on in-house development of biotechnology-based products, as opposed to those in Vietnam where licensing and in-house product development were major alternatives for firms engaged in product development. This result needs to be qualified slightly: Malaysia's model for biotechnology development remains the same as what it followed in electronics and other successful sectoral initiatives – it relies on attracting foreign firms to set up subsidiaries on its soil and the knowledge spillovers that can accrue from this. Hence the emphasis on in-house development, whereas Vietnam is largely focusing on developing a local sector, which explains the licensing.

Table 7.8 below compares the kinds of innovation that firms in all four countries are engaged in: whether they are only new to the firm, new to the local market only, new to the regional market, or new to the world at large. The table shows that much of the innovations are localized (new to firm and new to the local markets). However, Vietnam and Malaysia both have fairly high scores of innovations that are new to the

Table 7.7   Sources of technology for product development in biotechnology firms in Malaysia and Vietnam

Sources of technology for product development	Vietnam	Malaysia
1. Licensing	0.75	0.13
2. Foreign subsidiaries	0.015	0.06
3. Own development	0.87	0.67
Total number of firms	66	76

Source: Authors' survey (2007).

regional market (21% and 26% respectively) as opposed to Kenya and Nigeria (5% and 6% respectively).

We also found that, whereas Malaysia and Vietnam have extensive sector-specific dedicated biotechnology policies for finance and risk attenuation that are also being implemented, Nigeria and Kenya only showed policy frameworks that significantly obfuscated the innovation needs of the sector and focused largely on the other regulatory aspects. Table 7.9 below sums up Kenya's agricultural biotechnology policy framework. This does not show structured emphasis on the major aspects of innovation in terms of both generic innovation policy requirements and dedicated incentives for biotechnology-based work.

Table 7.8   Nature of innovation products in all four countries

Product	Vietnam	Malaysia	Kenya	Nigeria
New to the firm	66%	28%	17%	19%
Local market	63%	48%	17%	14%
Regional market	21%	26%	05%	06%
World	09%	02%	02%	.01%

Source: Authors' survey (2006–7).

Table 7.9   Policies, laws, and regulations relevant to biotechnology in Kenya

Policy/law/ regulation	Administering authority	Influence on biotechnology	Operational mechanisms
The Science and Technology Act (biosafety guidelines)	National Council for Science and Technology (NCST)	Provides guidelines and regulations for biotechnology	Biotechnology product development through NBC
The Trademarks Act	Kenya Industrial Property Institute (KIPI)	Promotion of trade and service marks	Registration of trade and service marks
Industrial Property Act	KIPI	Promotes IPRs through patents	Patent documentation and licensing
Seeds and Plant Varieties Act	KEPHIS	Registration and certification of new seeds and varieties	Plant health inspection and certification
Copyrights Act	Attorney General's office	Promotes protection of copyrights	Registration of copyrights

(Continued)

Table 7.9   (Continued)

Policy/law/ regulation	Administering authority	Influence on biotechnology	Operational mechanisms
Food, Drugs and Chemical Substances Act	Ministry of Health	Protects the adulteration of food, drugs and chemicals	Sets standards for food, drugs and chemicals
Pest Control Products Act	Pest Control Products Board (PCPB)	Provides for containment of pests	Inspection of agricultural materials and products
The Standards Act	Kenya Bureau of Standards (KEBS)	Sets standards for quality, purity and labelling	Standards setting, verification and implementation of codes of practice (CoP)
WTO rules (TRIPS and SPS)	Ministry of Trade and Industry	Mutually supportive relationship between TRIPS and CBD for biotechnology and sustainable development	Identification of source of materials, traditional knowledge and trade issues
WIPO administered treaties	KIPI	Plant protection treaty Madrid protocol on Trademark	Patents protection Trademarks protection
Cartagena Protocol on Biosafety	National Environmental Management Authority (NEMA) and NCST	Ensures human, animal and environmental safety	Safe transfer, handling and use of living modified organisms
Environmental Management and Coordination Act (EMCA)	NEMA	Ensures environmental safety	Provides for environmental impact assessment (EIA)

Source: Compiled by the authors.

Kenya does have a national science, technology and innovation policy that has been discussed since 2006, but its focus on enforcing its provisions has been weak and may take a long time to show results.[9] Comparing this with Malaysia, for example, one finds that although biotechnology was prioritized as a lead sector only in 2000, the policy focus has been substantiated by the extensive state apparatus created to aid in creation of new biotechnology firms, novel finance mechanisms, scientific investment particularly aimed at the sector, and collaboration incentives.

7.6 The role of state capacity and policy vision

In the case of agriculture, a vision to achieve self-sufficiency in food grains and enhancing crop productivity have been the drivers of the Green Revolution among those countries who successfully achieved it, and are following it up with biotechnology-based research programmes. While all states formulate goals of biotechnology-based development to deal with local issues of food security, the crux lies in the enactment of institutions and funding to see them through. Lack of state vision is for most part responsible for the poorly developed innovation capacity and reflects in several ways, including:

(a) Duplication of results across agencies set out to perform various institutional mandates;
(b) duplication of research efforts and wastage of scarce resources available for innovation;
(c) a lack of suitable reward structures that promote performance, and hence a resulting 'market for lemons' (Akerlof,), where researchers who are interested in serious R&D are forced to leave the country and over a period of time, only second-best graduates are available for recruitment in national agencies for research and management, and
(d) finally, the poor policy convergence between innovation investment and local disease and food challenges.

Of the four countries studied, we found immense differences in the way states had put in place different incentive regimes including those dealing with intellectual property and trade-related issues for agricultural biotechnology. It is precisely state vision and policy capacity that comes to the fore when the experience of Malaysia and Vietnam is compared to that of Nigeria and Kenya. Although biotechnology was prioritized only in 2000 in Malaysia, accelerated growth of the sector is being ensured through a variety of generic and dedicated biotechnology policies and policymaking has proceeded on an informed and targeted basis. The East Asian economies have mastered the art of championing policy visions of technology-led development in key sectors, and are aware that the state needs create a level playing field where it works in tandem with all other actors in the system.

The four aspects that stand out are:

1. The Malaysian government has clear projections of the growth of the sector up until 2020 and what level of employment and livelihood opportunities it should generate, along with the level of technological mastery that needs to be achieved.
2. A review of the spate of human skills in the country, how much more is required to support the expansion of the plans, and programmes to enable this human skills base are all underway.
3. A range of dedicated biotechnology policies that cater to finance and clustering needs of the sector, apart from those that provide incentives for collaboration have been put into place, and are being constantly revised based on interim outcomes.
4. The state is a master of championing economic development by acting along with all other actors in the system.

Vietnam, similarly, demonstrates extensive state policy capacity for a very latecomer, as demonstrated also through the example below.

State vision example 1: Vietnam and GM rice.

Despite the general caution exercised by latecomer countries towards transgenic crops, Vietnam was one of the earliest countries to plant Bt rice, which is a transgenic variety with a trait that protects harvests from the stem borer – the most common rice pest in Asian countries. Vietnamese rice production has grown by 4% over the decade between 1995–2005 and was accompanied by a significant rise in exports, however, its decision to adopt GM rice in 2008 is expected to be accompanied by huge gains in the coming years (see the results of the projection model in Maramil and Norton, 2006).

Significantly, what we found of relevance to the point under discussion is that Vietnam is one of the few countries that has embarked upon Bt rice production in Asia (apart from Philippines), and our interviews reveal that:

(a) The state has played a main role in creating trust, ensuring information asymmetry among all actors in the agricultural system of innovation and in engaging stakeholders to promote the Bt rice concept from the onset.

(b) The country has put in place significant public and private research investments to promote the production and dissemination of Bt rice technologies.

(c) These are accompanied by state competence in monitoring the environmental and other effects of Bt rice.

The interviews also reveal significant state planning in terms of projected gains and upfront investments that needed to be made by the state in order to ensure that these gains accrue. This, as we note in Chapter 8, is the imperative of state vision. This is again very different from the slow and very cautious progress that one observes in the case of Bt maize in Kenya (see also, Mugo et al., 2005)[10] where the state demonstrates significant inertia in championing the cause and realising the potential of the scientific progress being made.

State vision example 2: Malaysia and Oil Palm

Malaysia demonstrated vision in the way in which state policy exploited the synergy between large estate farms with the preponderance of labour in smallholder farms. Large estate farms are normally associated with higher productivity, they tend to adopt efficient technologies and are better positioned to use own savings or the credit market for investment. Innovation in seeds, as well as farm machinery has made it possible for large acreages of land to be profitably cultivated in order to take advantage of scale economies to lower labour costs and realize higher farm outputs. Malaysia deliberately put in place mechanisms that fostered synergy between large estates and smallholders to not just be co-located but to co-produce using institutional and organizational incentives and government policies. The productivity of smallholders is usually low, particularly those employing low technology traditional systems. In Nigeria, a large portion of agricultural ownership, more than

Table 7.10 Ownership of oil plantation in Malaysia and Nigeria

Ownership	Nigeria (%)	Malaysia (%)
Private large estate	12.7	60
Government/state schemes	7.0	30
Smallholdings	80.3*	10

Source: Authors' calculations from available secondary data (2007); Chandran (MPOA 2004).
Note: *mostly wild groves.

90% are small-scale oil palm farmers. In Table 7.10 below the structure of ownership reflects the different structure of farm ownership.

7.7   Agricultural innovation capacity, food security, and poverty reduction: A look at the broader implications

At the heart of efforts to reduce poverty in very latecomer countries is the well-known problem of rural smallholder farmers and also the real challenges presented by the expanding urban poor. The large majority of the chronic poor live in the rural areas of the very latecomer countries of SSA and South Asia; this dubious distinction will increasingly be shared by the poor in cities in the developing world (see Figure 7.5). In other words, attenuating mass poverty will not be possible without the application of new methods. These will of necessity have to include scientific application to farming as well as new institutional mechanisms to raise the income of poor households and any improvement in their incomes will depend on generating more and better jobs in rural areas.

Recent cross-country studies (Hazell et al., 2007) show that growth in agricultural productivity is strongly related to poverty reduction since in any case other opportunities for employment are few or non-existent in rural very latecomer countries. Quite clearly state action will be required to attenuate price variability in markets, to take some lead in promoting non-state activities and we see in the cases of Vietnam to re-orient actors (from state-led to private) and in Malaysia to promote synergetic relationships between public and private actors. Fragmented and inchoate markets such as we have in very latecomer countries require deep-going institutional innovation for them to provide functions which markets are poorly developed to perform in poor countries. In the framework of this book, latecomer development needs to develop their innovation capacities in order to deal with the issues of poverty and food security.

In this respect, we recall that iinnovation capacity relates to both human capabilities as well as research and knowledge infrastructure. We documented the strong relationship between technical skills and institutional capability, on the one hand, and organizational performance, on the other. For instance, the proportion of in-house university graduates and technical skills is a proxy of organizational capability of the university or RDI. To make matters more critical for latecomers, the intensified competitive global market environment is making a greater demand not only for more skilled managers and scientists, but also a wider range of skills are being required. The skills market faces persistent market failure particularly in latecomer countries and the nature of the market failure is not limited to skills but include access to information, finance, and technology markets. Small farms and resource-constrained farms are differentially penalized by information asymmetry, poor access to investment, and meagre working capital. In other words, latecomers face not just the problem of supply to local markets at affordable prices but a major hurdle in competing in the global market. Let us consider what these translate to in food production and poverty reduction.

In the period of 1980–2000, food production per capita figures were as follows: in Asia it grew by 2.3% per year, in Latin America by 0.9% and declined in Tropical Africa by 0.01%. Extreme poverty 1990–2001: People living on US$1 or less per day declined by 25 % in South Asia and 50% in East Asia and grew by 4% in Sub-Saharan Africa; and more recent figures are shown in Figure 7.3 Regarding the MDG goal of halving poverty by 2015, all regions are on-track except Sub-Saharan Africa; this is closely related to the poor record of slow improvement in Sub-Saharan African agriculture and concomitantly rising poverty and hunger. For instance, of the 85% of poor that live in rural areas, 75% are poor households' small farmers, 50% of small farm households are undernourished, see Figure 7.4.

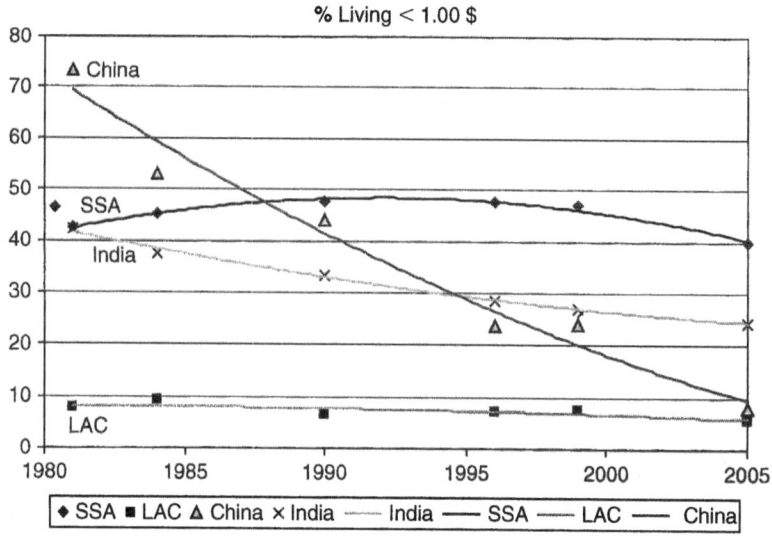

Figure 7.3 Percent of people living under US$1.00 per day
Source: Plotted from World Bank Poverty Data (2008).

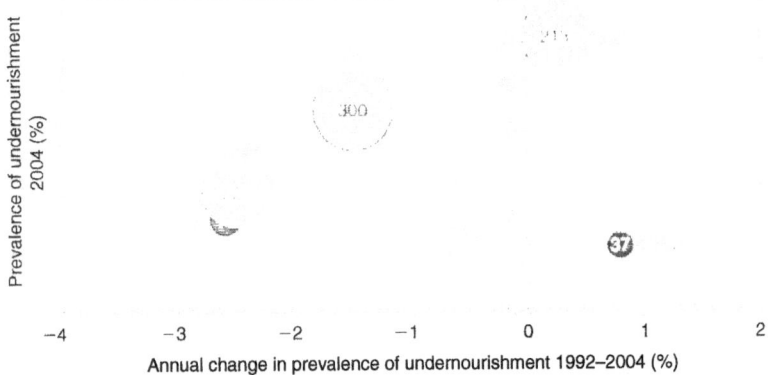

Figure 7.4 Prevalence of undernourishment in developing countries
Source: FAO 2006 and World Bank 2007.

The lack of progress to improved agriculture and towards providing basic subsistence or improved nutrition, tends to be related to unequal opportunities to earn income, and a small and fragmented private sector that should provide inputs and markets. Progress in reducing global poverty has been due largely to China, see figure (from 70% < US$1 per day in

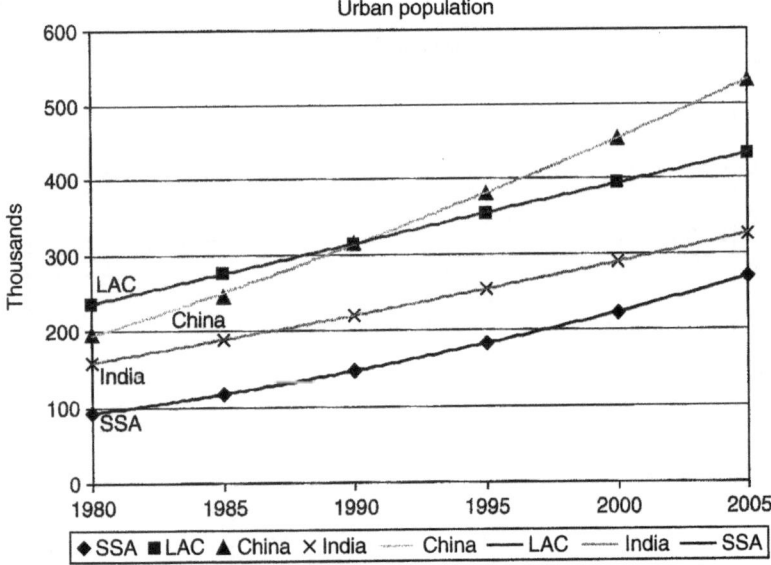

Figure 7.5 Population of people living under US$1.00 per day
Source: Plotted from World Bank Poverty Data (2008).

1982 to 10% < US$1 per day in 2005). In the same vein, urbanization is rising rapidly in the poorest latecomers putting real pressure on food and shelter conditions (see Figure 7.5). The resultant effect of these is the prevalence of undernourishment in latecomers that remains severe in SSA and South Asia. The need to resort to new technologies is urgent and real.

7.8 Summing up

This chapter sums up the basic conclusions of the last six chapters. The broad outlines re-examine the key lessons and perspectives that were thrown up from the different case studies analysed within a sectoral system of innovation. We coded the broad capabilities in terms of the *innovation capacity* of latecomer economies. The country experiences in accumulating biotechnology capacity show systematically that the nature of development and catching up is a complex landscape of both hope and distress particularly for food security and economic development among latecomers. We conclude by linking micro-evidence with broad national and regional examples and to show that scientific and technological capabilities coded as innovation capacity, can be applied to the resolution of food supply problems and poverty reduction. In the next chapter we sketch out policy lessons for very latecomer countries as well as donors, an urgent assignment in a time of an unprecedented global crisis that leaves large swathes of regions in Africa and Asia hungry and malnourished.

Appendix

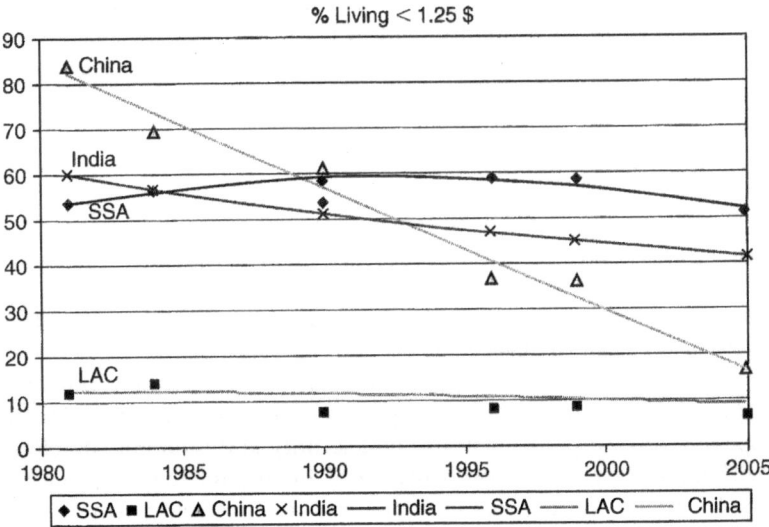

Figure 7.6 Proportion of Population Living below Poverty (1980–2005)

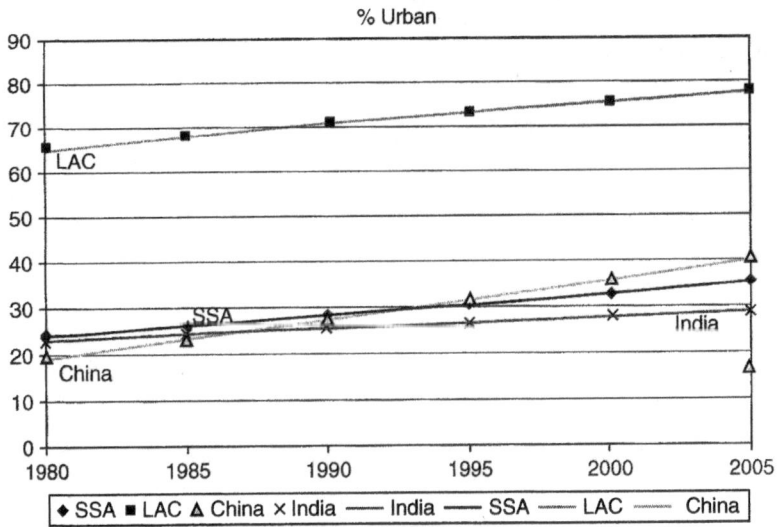

Figure 7.7 Regional Urbanization Trend (1980–2005)

8
Policy Insights and Recommendations

8.1 Introduction

In this short and final chapter, our aim is not to lay out detailed policy recommendations for different contexts and actors but rather to call attention to the broad underlying issues that can be used to frame policy. Evidently, the challenges related to catching up in biotechnology application in order to resolve the food crisis and poverty in those latecomer examined in this book are extremely complex and, sometimes, rather frightening. For instance, as we pointed out in the previous chapter, latecomers hold a mere 3% of all patents worldwide, 80% of these are in the hands of corporations and individuals in the highly industrialized countries and the world's top five biotechnology firms control over 95% of gene transfer patents.[1] Our data shows the evident connection between poverty and hunger: regions and groups that record the least income are the most severely penalized measured by those that suffer the most malnutrition and under-five mortality. These regions and/or groups tend also to be the most land-constrained and to have minimal access to education. The least progress in attaining the MDGs is recorded by the poorest regions of SSA and South Asia. Yet the vicious cycle of poverty equally forecloses the imperative of investment in what is most critical – the immediate need to commit resources to building innovation capacity through investment in scientific and engineering manpower and the construction of laboratory and industrial facilities in order to focus on urgent problems local food supply and disease. In other words, we need to address urgently the challenges of institutions, infrastructure, and human resources that lead to exclusion and deprivation and, secondly, to break the cycle of a lack of access to credit in poor countries. Donor focus will attenuate temporary suffering; it will never break the gridlock of persistent poverty. Policies should therefore focus on investments to redress the poverty of countries and households by making them self-sufficient in food production through the application of new technologies such as biotechnology.

In sum, we agree in large part with much of the findings of the various United Nations organizations and the more recent evidence and prescriptions of IAASTD (2008). *First*, as we have repeatedly observed in the book, the perspective of systems of innovation show us the intricate and complex linkage of agriculture, nutrition, and health and, for this reason, income, per capita yield of crops output, and scientific improvements to local crops produced by the majority of small farmers will be key to solving the poverty problem. *Second*, as the majority of smallholder farmers are on average the poorest, they are excluded from the credit market which makes larger agricultural outputs possible. The markets in these countries are fragmented and largely informal. Policy should be directed at *clustering the thousands of small farmers within geographic locations*. Clustering makes it possible for isolated small farms to act as one big farm within an institutional framework of say joint production and cooperatives in their different contextual modes. This recommendation derives from the idea that small organizations are not necessarily deprived due to their size but rather largely due to their isolation (Oyelaran-Oyeyinka and McCormick, 2007). *Third*, the nature and modes of intervention should address the poverty of knowledge of policymaking at the same time as it tries to build local innovation capacity. Poor state intervention is the result of two serious challenges in latecomer development: the complexity of the underlying problems and the perennial lack of funds.

8.2 Delineating development across continents: Policy insights for agricultural innovation in Asia and Africa

Whether the Green Revolution that was so successful in the Asian and Latin American hemisphere can be replicated elsewhere, and what the pre-conditions of that process may be is an issue that has been debated at length in the literature. This book and its analysis contributes to this debate by showing that without the institutional rubric of the Green Revolution, achieving the biotech and genomics-induced Gene Revolution will only be a distant reality for latecomer countries because the latter build extensively on the institutional infrastructure of the Green Revolution. This finding is highly relevant and feeds into the views of the United Nations and several other multilateral agencies, which have pointed out that the cost of leaving out some countries from the biotechnology-based Gene Revolution, will be higher than the cost of ensuring such technological capacity is built.

While the policies all have the aim of achieving food self-sufficiency and comparative advantage, what remains is that while formulating action plans and strategies and coordinating these through various organizational arrangements, the lofty ambitions often become unrecognizably blurred. This, as we have identified, lies within the organizational competence of the agencies themselves as well as the capacity of the state to drive the agenda and not just identify it, as one worthy of pursuance. Driving the agenda, however, is a difficult and costly task. It calls for accountability, monitoring and evaluation, and team building efforts between the various actors in the sectoral system of innovation, including often creating complete missing sets of actors. This is precisely the state vision and policy capacity that emerges when the experience of Malaysia and Vietnam is compared to that of Nigeria and Kenya. Although biotechnology was prioritized in Malaysia only in 2000, accelerated growth of the sector is being ensured through a variety of generic and dedicated biotechnology policies and policymaking has proceeded on an informed and targeted basis. The East Asian economies have mastered the art of championing policy visions of technology-led development in key sectors, and are aware that the state needs create a level playing field where it works in tandem with all other actors in the system. Vietnam, once again, makes for a highly interesting country comparison simply because it demonstrates extensive state vision and policy capacity despite its status as a least developed country.

Our country-level findings and comparative analysis show that it is precisely the countries that do not have the basic infrastructure to enable the Green Revolution that are lagging behind enormously in agricultural biotechnology. They are unable to benefit in a holistic sense from the several international projects that are ongoing within their local contexts as a result of the limited absorptive capacity of the sectoral innovation system.

Our comparison between African and Asian countries further reveals several issues of importance to policy. The Asian countries considered in this book have the fast movers' advantage because their basic agricultural innovation system that is a pre-requisite for the Green Revolution is well in place, and both Malaysia and Vietnam show adequate state capacity in enacting policies and coordinating them for the development of key sectors. The government's role in promoting research is critical and evidence from the surveys at the country level is not conclusive enough to point to the awareness of the policymakers in this regard. The state governments in Nigeria and Kenya, for example, are either not persuaded of the role of these organizations or are simply unaware of the institutional and system failure evident within their own contexts. Of the 210 organizations surveyed in Nigeria, only 40% have research funding sources from the government, and only 19% of them were of the opinion that government incentives played a role in product and process innovations.

In Africa especially, these can be reduced to two priorities: first, ensuring that the present advances in biotechnology and genomics are steered towards manipulating the content and composition of local crops such as cassava, sweet potato, cow pea, and millet. Second, ensuring these initiatives focus not only on yield and pest resistance, but also on the nutritional value of the crops, which is within the reach of today's technologies. The nutritional value of the food is a priority for Africa, where more than one-third of its population is chronically undernourished.

Policy reform needs to begin with the agricultural innovation system and the focus needs to be on sector-specific agricultural policies that are designed to facilitate agricultural marketing, reduce agricultural production costs, and enhance agricultural product prices as incentives for increased agricultural production. Additionally:

(a) Macro and institutional policies for innovation as well as legal frameworks should complement sector-specific policies.
(b) Constraints to agricultural policy effectiveness are identified to include those of policy instability, policy inconsistencies and the narrow base of policy formulation, poor policy implementation, and a weak institutional framework for policy coordination.
(c) Local food security, self-sufficiency, and innovation capacity should be the primary objectives of the policies.
(d) Rational utilization of agricultural resources, the promotion of increased application of agricultural technology, and an improvement in the quality of rural life should be built into organizational mandates and policy agendas.
(e) International collaborations should be strategically structured to build capacity in biotechnology innovation in a systematic way.
(f) Secondary policy instruments that reduce risks and uncertainties of biotechnology-based innovation, as well as catering to manpower development need to be enacted and well coordinated with the underlying agricultural system of innovation.

(g) Unified national agricultural extension system needs to be prioritized and an agro-based entrepreneurship and industry needs to be
fostered.

8.3 Endogenous capacity for GM crops in Africa

Kenya's primary achievement with the transgenic sweet potato case
was to demonstrate the potential for collaboration between African
research institutes and the international private sector that incorporated capacity-building, intellectual property sharing, and the freedom
to develop new varieties. But this has since then not been duplicated
and the pressing question that needs an answer in the present context
is: Do African public-sector institutions have the capacity to bring out a
new GM crop that caters to its local needs?

If GM crops are to address food security in the long term, varieties
need to be developed that can cater more effectively to local needs.
Ideally, latecomer countries require GM crops that focus more on
output traits such as higher nutritional value, while a the same time
catering to certain input traits such as resistance to abiotic stresses – in
addition to the already available herbicide tolerance and insect resistance traits – in order to solve their food and nutrition problems. A closer
look at only those GM varieties that are in the research and development (R&D) pipeline in Europe (both private and public sector) shows
that varieties with mainly input traits are being released; whereas varieties that are dominated by input traits but have some output traits are
also slowly finding their way into the market. In addition, GM varieties
that contain abiotic stress factors and a higher content of 'functional'
ingredients are likely to be released only after 2011 (Lheureux et al.,
2003, pp. 23–4; Niang et al., 2004).

This means that the focus of research has to shift from those crops
that are perceived to have a commercial value by the private sector
to those that cater more to local needs in latecomer countries. This
remains doubtful, due to the innovation capacity constraints that we
have identified throughout the book, and specifically highlighted by
comparisons in the earlier sections of this chapter. This preliminary
relies on the basic infrastructure: namely, S&T infrastructure, a seed
development and distribution system (which is the most critical), and
an institutional capacity to create and monitor policy. Moving from
confined trials to pre-commercial trials requires the capacity to increase
seed amounts, and to provide larger areas for testing purposes. And
more importantly, products need to be made available to farmers only

after *commercial release*, through privately or publicly owned seed companies or other institutional mechanisms. These mechanisms, however, are unavailable.

We also find, similar to the conclusions made in Niang et al. (2004), that the lack of enforceable guidelines and mechanisms for GM crops in African countries means that the apparatus required for approving and conducting commercial trials is missing in its entirely. Furthermore, our surveys clearly show that the public sector does not have the capacity to monitor confined trials for safety and efficacy. The state also does not have either the capacity to guarantee extension services, seed supplies, and the ability to evaluate product performance on a large scale, or to include experiments designed specifically for safety evaluation. In short, the public-sector performance, capacity, and incentives all appear not to be synchronized to perform these kinds of demanding function for the present.

In order to achieve the 'Gene Revolution' in Africa, addition to dealing with these two factors, African governments and policymakers have to address a range of policy reform issues, including:

(a) Creating a large base of skilled manpower in Africa, through the strengthening and reform of the university education system.
(b) The importance of local food security as a major policy focus: for example, in Nigeria only half of the total arable land is being used for local agriculture and a further half of the local agriculture is directed towards cash-crop production.
(c) The importance of liaising with donors and international agencies to build capacity in areas of relevance to local needs: Most projects are for field trials and biosafety. This is a disturbing trend and tends to take away precious focus from building capacity *per se*. This needs to be addressed by national policymakers in a predominant way.
(d) The imperative is to balance external demands through the lens of local science, technology, and innovation aspirations. As simple as it may sound, local priorities need to be set locally, with the involvement of local stakeholders and bearing stock of local strengths. In the absence of this, the knowledge basis required for the Gene Revolution cannot be established.

Notes

Preface

1. Oyelaran-Oyeyinka and Gehl-Sampath, P., 'Latecomer Development: States, Knowledge and Economic Catch-up, Routledge, September 2009; Gehl Sampath, P., Unhealthy Divide, Local Capacity for Disease of the Poor, 2010; Oyelaran-Oyeyinka, Banji and Rajah Rasiah (2009), *Uneven Paths of Development: Innovation and Learning in Asia and Africa*, Edward Elgar Publishing, United Kingdom.

1 Agricultural Biotechnology Innovation Capacity and Economic Development

1. The innovation system is defined as the set of organizations (e.g., firms, universities, and public laboratories) and their linkages through which innovation processes develop. For an in-depth analysis of innovation systems in developed countries, see OECD (2000).
2. In addition the author suggested five 'accelerators' to get 'get agriculture moving'. Much of Mosher's five 'essentials' presaged the key issues being put forward by current scholars of 'pro-poor agricultural growth'. These accelerators are: (1) education for development; (2) production credit; (3) collective action by farmers; (4) improving and expanding agricultural land (through good husbandry and adoption of appropriate technologies), and (5) national planning for agricultural development with emphasis on enlightened public policies that support the agricultural sector, see Thompson et al. (2007).
3. The notion of scientific and technological convergence is discussed in the next section.
4. From 2005 till mid-2008, maize prices almost tripled, whereas wheat prices and rice prices increased by 127% and 170% respectively. At the same time, palm oil prices and soybean oil prices increased by 200% and 192% respectively.
5. The role of knowledge infrastructures and historical investments in universities and industrial and agricultural public research in Germany, Japan, the US, and recently in Taiwan Province of China and the Republic of Korea, have been well documented, see Mowery (2005).
6. Banji Oyelaran-Oyeyinka and Padmashree Gehl Sampath, Latecomer Development: Innovation and Knowledge for Economic Catch-Up (Routledge, 2009).
7. Edquist (2001) identifies nine broad similarities, namely: All SI approaches place innovation at the centre of activities; innovation processes are evolutionary in nature; all reject the concept of optimality and emphasize diversity and variety; they take innovation as an interactive learning process; they stress the interdependence between organizational actors, and they affirm that innovation generally occurs within an institutional context.

8. See Bell and Pavitt (1993) who make the distinction between technological capacity and capabilities; also the Hall and Dijkman (2006) articulation of innovation capacity in agriculture biotechnology.
9. R&D, however, it is defined, is not only an avenue for economic and social diversification (new products and processes) but also helps build scientific and technical competencies.
10. The Precautionary Principle was defined at the 1992 Rio Declaration on Environment and Development (UNCED, 1992) in a much broader way, and applies to a range of issues apart from agricultural biotechnology and GM crops. It was defined to be: 'Where there are threats of serious or irreversible environmental damage, lack of full scientific certainty shall not be used as a reason for postponing cost-effective measures to prevent environmental degradation.'
11. See Edquist and Hommen (1998b) for a discussion of the subject.
12. This question naturally excludes the capacity of large multinational conglomerates operating in latecomer countries.

2   Sectoral Systems for Agricultural Biotechnology

1. This definition of an innovation system draws upon the work of Nelson and Winter (1982); Lundvall (1988); Freeman (1988), among others.
2. This and the next paragraph is based on Oyeyinka and Rasiah (2009).
3. See in this context, the analysis in Gehl Sampath, 2009.
4. Today's modern industrial societies evolved from a system of personal to impersonal exchange that is now build on complex institutions and highly knowledge specialized (North, 2005; Mokyr, 2002). 'As increased specialization occurred with the growth of markets, individuals exchanged increased specialized knowledge at the expense of less "general" knowledge'(North, 2005, p. 122).
5. Amsden (1977) cites Allyn Young's clarification of what an 'extent of market' is.
6. She illustrates: 'Assume market 100 economic units in market A with an income of US$10000. Assume 1000 economic units in market B each with an income of US$1000 ... purchasing power in the two markets is equal, obviously a market of type A is a better candidate than a market type B for the absorption of non-essential goods with high unit costs, irrespective of how great increasing returns may be, and hence relatively high prices' (Amsden, 1977, p. 218).
7. Biggs and Shah (2006).
8. David and Foray (2000) similarly identify four different communities, namely, the scientific community which is concerned with the 'capture, storage, analysis, and integration of experimental and observation data' (p. 7). Another is a community of programmers involved with the open source movement. They create new knowledge and information with no profit motive. The third is the business community whose *modus operandi* is cooperative and organizational such as education and research consortia and networks. Another science-driven community is the health and medical profession whose practices are being increasingly science-driven.
9. Several formal channels of transfer of technology have been identified in the literature including but not limited to: migration of skills, licensing, trade

and foreign direct investment (FDI), turnkey projects, technical consultancy, capital goods imports, and joint-venture agreements.

10. Technological capability derive from both internal and external knowledge and as such firms require an absorptive capacity defined by Cohen and Levinthal (1990) as the ability of a firm to recognize new information, assimilate it, and apply it for commercializing inventions. They argue that successful exploitation of basic scientific outputs require firms to continuously learn from external sources.

11. George et al. (2002) analyse 2457 alliances undertaken by 147 biotechnology firms, in order to arrive upon these findings. However, as the authors themselves note, it is not clear as to how much such collaborations influence (or enhance) the financial performance of the firms.

12. See Shartinger et al. (2002) and their references to different shades of the issue: academic research and biotechnology spin-offs from universities and PRIs (Fontes, 2004); transfer between university research and industry, (Lee and Win, 2004); on the overall impact of university research on industrial production (Jaffe, 1989; Anselin et al., 1997), and personnel mobility, (Bania et al., 1992; Almeida and Kogut, 1997).

13. Lubit (2001) identifies four categories of tacit knowledge, namely, (a) hard to pin down skills – 'know-how", (b) mental models, which show us how the world is constructed, (c) ways of approaching problems, and (d) organizational routines. According to him, 'The word skill implies tacit knowledge which range from the ability to swing golf balls to the dexterity of handling cells in a biology lab, all which are hard to explain in words.'

14. According to Lubit (2001; p. 167), 'Routines solidify as standard operating procedures and roles are developed and enforced. Routines include ways of producing things, ways of hiring and firing personnel, ways of handling inventory, decision-making procedures, advertising policy, and R&D procedures.'

15. Oyeyinka and Gehl Sampath, (2009); Oyeyinka and Rasiah (2009); Gehl Sampath (2009).

16. We have tested these inter-relationships using empirical data in earlier work, see for example, Gehl Sampath and Oyeyinka (2009), and Oyeyinka and Gehl Sampath (2009).

17. For instance Korean government efforts were directed 'to help the private sector accelerate technological efforts'. This effort focused on three dimensions namely: to induce the private sector to invest in technological efforts by creating a market for innovative products (*the demand side of technology*); to help them enhance R&D activities and acquire technological capabilities (*supply side of technology*); while the third component focused on efforts to 'provide an *effective linkage between the demand and the supply sides*, making technological efforts feasible and less risky and costly' (Kim, Lee, and Lee, 1987, p. 282).

18. Nelson and Sampat (2001) cite North's (1990) re-articulation of the 'institutional obstructionist' notion of economic backwardness as being responsible for the failure of poorly performing economies to adopt productive technologies.

19. Evans (1995) presents a detailed theoretical and empirical discussion.

20. As Cumings (1999: p. 64) observed 'Johnson takes the importance of technology to Japanese growth seriously ... MITI rode hard on the acquisition of new technologies and how at critical turning points MITI directed advanced

technologies and cheap finance to new industrial sectors – especially the shift to heavy and chemical industries in the 1950s'.
21. See Chang (1999; 1994) for details.
22. Countries have other incentives such as those in German law confer ownership of patents on individual researchers thereby they maintain full IPRs over inventions (Giesecke, 2000).
23. Chang (1999) and Vartianen (1999).

4 Vietnam Biotechnology: Building Local Capacity

1. The Resolution is dated on 11 March 1994.
2. We employ by maximum likelihood a probit model of new product and/or process development. We also consider a dynamic panel data type 1 tobit model that studies the persistence of R&D investments. The model accounts for unobserved firm heterogeneity through individual effects that are assumed to be random. We use the maximum likelihood (ML) estimator and solve the so-called 'initial conditions' problem using Wooldridge's (2005) 'simple solutions' that consist in writing the individual effects, in each period, as a linear function of the explanatory variable and the initial conditions.
3. Firm's total sales are also available for each year of the period 2001–5 but no included in the analysis because of too many missing values.
4. Table 4.4 reports 12-point Gauss-Hermite ML estimation results of the dynamic type 1 tobit.
5. The estimated annual expenditure in practice is approximately US$15 million.
6. See tables in Appendix.
7. In spite of the myriad of legislations, Vietnam still lacks laws and regulations regarding biosafety and traditional knowledge, which are important for the development and operation of biosafety sector.
8. In our case, the vector x_{it} actually consists of one single variable, hence a scalar.
9. The vector x_{it} in order to be included in equation (2), must be sufficiently time-variant, otherwise a collinearity problem will arise.

5 Nigeria as a Very Late Follower in Agricultural Biotechnology

1. Other scholars who have worked on this issue in the country make similar observations; see for example Adeoti (2005).
2. As of 2004 according to the National Economic Intelligence Committee, about 0.6% (250,000 persons) out of the economically active population of about 40 million constitute the scientific and engineering manpower of Nigeria, when compared with about 10% in China, 11% in South Korea, 12% in Malaysia, 15% in Japan and about 25% in Russia, the US, Germany, the UK, and France.
3. To just state an prominent example, a total take-off enrolment of 210 was recorded for all disciplines in 1948 for University College Ibadan, the Premier University. This went up to 23,000 in the six universities in 1962. By 1996, the total student enrolment figure had risen to 234,581 for 37 universities

and by March 2002 it shot up in excess of 526,780. This was supported by little investments into expanding the quality and quantity of educational and research infrastructure at the college.

4. This result once again confirms earlier work on agricultural biotechnology in Nigeria (see for example, Adeoti, 2005).

5. Our 2003–4 survey was much broader and looked at biotechnology across both the pharmaceutical and agricultural sectors. These survey results showed that in fact, Nigerian PRIs were more advanced in using biotechnology techniques for drug and vaccine production (in a relative sense) (see Oyelaran-Oyeyinka et al., 2005; Oyeyinka and Gehl Sampath, 2007).

6. Alhassan (2001) documents that of the 14 laboratories using tissue culture, only five were considered to be functional on the basis of stable power supply and possession of the minimal required equipment.

7. As Jolly (1997) correctly observed, technologies and for that products and process inventions fail not so much for the skills of the inventor and the lack of market but because no one promotes or get sufficiently interested in them.

8. The NBA was only proposed and still not constituted at the time of the survey.

6 Kenya's Incipient Innovation Capacity in Biotechnology

1. See also in this context, Gehl Sampath, The Unhealthy Divide: Local Capacity for Diseases of the Poor, 2010.

2. We note here that in the Nigerian context, the education level of the researchers is not the main reason impeding progress into other more challenging biotechnologies, whereas in the Kenyan case, this is an important reason.

3. Eaton et al. (2007) arrive upon a similar conclusion in their comparison of plant variety protection regimes in several countries, Kenya being one of them.

7 Comparative Analysis of Innovation Capacity in Latecomer Countries

1. See Oyelaran-Oyeyinka and Gehl Sampath, Latecomer Development: Innovation and Knowledge for Economic Catch-up, forthcoming Routledge, 2009, for an analysis of 75 countries including the four considered up close here.

2. Total public spending as a percentage of agricultural output (AgGDP) is a common research investment indicator that helps to place a country's agricultural R&D spending in an internationally comparable context.

3. The other countries are South Africa, and Kenya, Sudan, and Ethiopia – all three located in East Africa.

4. At the turn of the twentieth century, for example, Lord Leverhulme approached the British colonial governments in West Africa for land concessions to develop oil palm and was turned down.

5. One of such is a dedicated fund used for targeted research areas in Malaysia and the establishment of Round Table Sustainable Palm Oil (RSPO) which has all the players in the industry as members.

6. We make a distinction between donor programmes and international link-ages, which is explained at length in the next section.
7. See www.upov.org, accessed on 13 April 2009.
8. See International Assessment of Agriculture Knowledge, Science, and Technology for Development (2008), Issues in Brief, www.islandpress.org/iaastd.
9. A National Science, Technology and Innovation Strategy has now been in force since last year.
10. See also, Harsh and Smith (2007) for an interesting perspective on govern-ance issues in Kenya in the case of tissue culture bananas.

8 Policy Insights and Recommendations

1. See International Assessment of Agriculture Knowledge, Science and Technology for Development (2008). Issues in Brief: www.islandpress.org/iaastd.

References

Abler, D. G. and V. A. Sukhatme (2006). 'The "Efficient But Poor" Hypothesis', *Review of Agricultural Economics* 28: 338–43.

ACTS (2005). 'Introduction to Biotechnology and Biosafety: A Policy Analysis Training Course Designed With Reference to the High-Level African Panel on Modern Biotechnology of The African Union (AU) and the New Partnership for Africa's Development (NEPAD)', 24–28 October 2005.

Akerlof, George A. (1970). The *Market for "Lemons"*: Quality Uncertainty and the *Market* Mechanism', *Quarterly Journal of Economics* 84(3): 488–500.

Alhassan, W. S. (2001). *The Status of Agricultural Biotechnology in Selected West and Central African Countries*, IITA, Ibadan.

Amsden, A. H. (1989). *Asia's Next Giant: South Korea and Late Industrialization.* New York: Oxford University Press.

Amsden, A. H. and W.-w. Chu (2003). *Beyond Late Development: Taiwan's Upgrading Policies*. Cambridge, MA: MIT Press.

Annual Report, Malaysian Biotechnology Corporation (BiotechCorp), 2006.

Bell, M. and K. Pavitt (1993). 'Technological Accumulation and Industrial Growth', *Industrial and Corporate Change*, 2 (2): 157–211.

Bell, M. and K. Pavitt (1997) 'Technological Accumulation and Industrial Growth: Contrasts between Developed and Developing Countries', in D. Archibugi and J. Michie (eds), *Technology, Globalisation and Economic Performance*; Cambridge: Cambridge University Press, pp. 83–137.

Biomedical Science (2002). 'Can Money Turn Singapore Into a Biotech Juggernaut? Normile', *Science*, 30 August: 1470.

BiotekAfrica News Issue No. 1, 2003. 'A topical News brief from ABSF and KBIC', Nairobi, Kenya.

Blakeney, M. (1999). 'The International Framework of Access to Plant Genetic Resources', in *Perspectives on Intellectual Property: Intellectual Property Aspects of Ethono-Biology*. London: Sweet and Maxwell.

Bolo, M. (2004). 'Science and Governance of Modern Biotechnology in sub-Saharan Africa: The Kenyan Case Study', unpublished manuscript.

Bolo, M. (2005). *Agricultural Biotechnology and Wealth Creation: Prospects for the Youth. The Case of Tree Biotechnology Project in Kenya*, in CTA and ATPS (2005), The African Regional Youth Congress and Exposition on Youth Employment and Wealth Creation: Opportunities in Agriculture, Science and Technology and Youth Leadership for HIV/AIDS Prevention. Proceedings report of the youth congress held on 20–23 June 2005 in Nairobi, Kenya.

Bokanga, A. and E. Otoo (1991). 'Cassava Based Food: How Safe are They?', in K. A. Taiwo, Utilization Potentials of Cassava in Nigeria: The Domestic and Industrial products. *Food Reviews International*, 22: 29–42.

Bougrain, F. and B. Haudeville (2002). 'Innovation, Collaboration and SMEs Internal Research Capacities', *Research Policy*, 31: 735–47.

Brown, J. and P. Duguid (1991). 'Organizational Learning and Communities-of-Practice: Toward a Unified View of Working, Learning, and Innovation', *Organization Science*, 2, 1: 40–57.

Bui Thi Huong (2005). *Vietnam Biotechnology 2005*, GAIN Report Number: VN5050.

Ca, Tran Ngoc (2003). 'International Cooperation in Science and Technology: Some Critical Issues for Vietnam', paper presented at MOST-IDRC roundtable on international cooperation in science and technology. Hanoi. October.

Cao Minh Quang (2006). *Vietnam Pharmaceuticals Industry – Opportunities and Challenges*. General Department of Pharmaceuticals Management. Hanoi.

Chandler Jr., A. (1997). *Scale and Scope: The Dynamics of Industrial Capitalism*. Harvard: Belknap.

Chang, H-J. and A. Cheema (2001). *Conditions for Successful Technology Policy in Developing Countries*, United Nations University, Institute for New Technologies, S. S. a. I, Learning Rents, Maastricht, The Netherlands, 2001–8.

Chi, Phan Van (2004). 'Current Status of Bioscience/Biotech in Vietnam', paper presented at Canada-Vietnam Roundtable on international cooperation in biotechnology. Ottawa. November.

Cohen, W. M. and D. A. Levinthal (1990). 'Absorptive Capacity: A New Perspective on Learning and Innovation', in *Administrative Science Quarterly*, Vol. 35, No. 1, Special Issue: Technology, Organizations, and Innovation (March) pp. 128–52.

Clark, C., J. Mugabe, and J. Smith (2005). *Governing Agricultural Biotechnology in Africa: Building Public Confidence and Capacity for Policy-making*. ACTS Press. Nairobi, Kenya.

Clark, N. and C. Juma (1991). *Biotechnology for Sustainable Development: Policy Options for Developing Countries*. Nairobi: ACTS Press.

Cook, M. (1993) 2nd ed. *Levels of Personality*. New York: Cassell.

David, P. and D. Foray (2002). 'An Introduction to the Economy of Knowledge Society',. *International Social Science Journal*, 54 (171): 9–23.

Dasgupta, P. and P. David (1994). *Toward a New Economics of Science* Vol. 23, 1994: 487–521 Cambridge University, Cambridge, UK.

De Koning, M. (1999). 'Biodiversity Prospecting and the Equitable Remuneration of Ethno-biological Knowledge; Reconciling Industry and Indigenous Interest', in *Perspectives on Intellectual Property: Intellectual Property Aspects of Ethnobiology*. London: Sweet & Maxwell.

Department of Biotechnology, Ministry of Science and Technology, Government of India, Annual Report 2004–5.

Department of Biotechnology, Ministry of Science and Technology, Government of India, Annual Report 2005–6.

Dosi, G. (1988b). 'Sources, Procedures, and Microeconomic Effects of Innovation', in *Journal of Economic Literature*, Vol. 26, Issue 3, pp. 1124–71.

Edquist, C. (ed.) (1997). 'Systems of Innovation Approach – Their Emergence and Characteristics', in *Systems of Innovation: Technologies, Institutions and Organizations*. Pinter, London.

Edquist, C. (2001). 'Innovation Policy – A Systemic Approach', in *The Globalizing Learning Economy*. D. Archibugi and B.-A. Lundvall, (eds) Oxford: Oxford University Press.

Edquist, C. and F. Texier (1996). 'The Growth Pattern of Swedish Industry 1975–1991', in Osmo Kuusi (ed.) *Innovation Systems and Competitiveness*,

Taloustieto Oy in Collaboration with ETLA (The Research Institute of the Finnish Economy) and VATT (The Government Institute for Economic Research).

Derek Eaton, Robb Tripp, and Nils Louwaars, 'The Effects of Strengthened IPR Regimes on the Plant Breeding Sector in Developing Countries', paper presented at the International Association of Agricultural Economists Conference, Gold Coast, Australia, August 12–18, 2006.

Economic Development Board Singapore, Biotechnology Report (2006).

Eighth Malaysia Plan, Government Printers (1999).

Eicher, Carl, K., Karim Maredia, and Idah Sithole-Niang (2006). 'Crop Biotechnology and the African Farmer', *Food Policy*, 31(6): 504–52.

Ernst, D., T. Ganiatsos, and L. Mytelka (eds). (1998). *Technological Capabilities and Export Success in Asia*. London: Routledge.

Evans, P. (1995). *Embedded Autonomy: States and Industrial Transformation*. Princeton, Princeton University Press.

Evenson, R. and D. Gollin (2003) 'Assessing the Impact of Green Revolution: 1960–1980', *Science*, 300: 758–62.

Chaturvedi, Sachin, 'Evolving a National System of Biotechnology Innovation: Some Evidence from Singapore', *Science, Technology Society*, (2005); 10: 105–27.

Fernandez-Cornejo and W. McBride (2000). *Genetically Engineered Crops for Pest Management in US-Farm Level Effect*. Department of Agriculture. Economic Report no 786 Washington DC.

Freeman, C. (1995). 'The "National System of Innovation" in Historical Perspective', *Cambridge Journal of Economics*, 19(1): 5–24.

GRAIN. 'Plant Variety Protection To Feed Africa?', December 1999, downloadable from www.grain.org/.

Gichuki, S. T. (2006). Current Status of Agricultural Biotechnology Application In Kenya, KARI Biotechnology Center, Kenyan Agriculture Resources Institute, Kenya.

Grace H. W. Wong. (2006). 'Chinese Biotech: The Need for Innovation and Higher Standards', *Nature Biotechnology*, 24, 221–2, doi:10.1038/nbt0 206–221.

Gehl Sampath, P. and G. Tarasofsky. (2002). *The Inter-Relationships Between Intellectual Property Rights and Conservation of Genetic Resources*. Ecologic – Institute for International and European Environmental Policy.

Gehl Sampath, Padmashree and Oyelaran-Oyeyinka, Banji (2009). 'Rough Road to the Market: Constrained Biotechnology Innovation and Entrepreneurship in Nigeria and Ghana', *Journal of International Development*, Spring 2009.

Gehl Sampath, P. (2010). *Unhealthy Divide: Local Capacity for Disease of the Poor*. Routledge.

Graff et al., (2005). 'Access to Intellectual Property is a Major Obstacle to Developing Transgenic Horticultural Crops', *Calif Agric*, 58: 120–6.

Global Agriculture Information Network (2005). *Nigeria Agricultural Biotechnology Reports* number N15014, 7pp.

Harsh, M. and J. Smith (2007). 'Technology, Governance and Place: Situating Biotechnology in Kenya', *Science and Public Policy*, 34(6): 251–60.

Idah Sithole-Niang, Joel Cohen, and Patricia Zambrano (2004). 'Putting GM Technologies to Work: Public Research Pipelines in Selected African Countries', African Journal of Biotechnology Vol. 3 (11), pp. 564–71, November 2004.

Ilori, M. O., Adeniyi, O. R. and Afilaka, S. O. O. (1994). 'Emerging Biotechnologies and their Potential Applications in the Production of Processing of Cocoa and Palm Produce in Nigeria', *Technovation* 14 (5): 287–94.

India *Economic News* (2002–3), vol. xii, Number 2, Winter.

Irefin, I. A., Ilori, M. O. and Solomon, B. O. (2005). 'Agricultural Biotechnology R & D and Innovations in Nigeria', *Int. J. Agricultural Resources, Governance and Ecology.* 4(1): 64–80.

International Service for the Acquisition of Agri-Biotech Applications (ISAAA), Global Knowledge Centre on Crop Biotechnology. Available at http://www.isaaa.org/kc/default.asp.

Ivanic, M. and W. Martin (2008). 'Ensuring Food Security', *International Monetary Fund, Finance and Development*, December 2008, Volume 45, Number 4, pp. 37–9.

Joseph W., Uyen Q., Halla Thorsteinsdóttir, Peter A. S. and Abdallah S. D. (2004). 'South Korean Biotechnology - a Rising Industrial and Scientific Powerhouse' in *Nature Biotechnology*, Volume 22, Supplement December 2004, DC 42–47.

Kaiser, R. and H. Prange (2004). 'Managing Diversity in a System of Multi-Level Governance: The Open Method of Coordination in Innovation Policy', *Journal of European Public Policy* 11 (2): 249–66.

Katz, Jorge (2004) 'University of Chile and Conicet, Argentina', paper presented at the World Bank workshop on the 'How to' of Technological Change and Adaptation for Faster Growth, February 16–17.

KEPHIS (2004). 'Ensuring Safe Handling, Testing and Utilization of Genetically Modified (GM) Crops in Kenya', a write up for publication as a supplement on Biotechnology and Biosafety in the East African Standard.

Kirea, S., V. O. Awuor, and S. S. Atsali (2003). 'Biotechnology Research and Development in Kenya: Policy Background', Study report on Product development Partnerships. Discussion Paper No. 03–9. ISNAR.

Kuta, D. D. (2004). 'GM Technology to Benefits Farmers in Nigeria', *AgBiotechNet* 6:120–6

Li Zhenzhen, Zhang Jiuchun, Wen Ke, Halla Thorsteinsdóttir, Uyen Quach, Peter A. Singer, and Abdallah S. Daar (2004). 'Health Biotechnology in China - Reawakening of a Giant', in *Nature Biotechnology*, Volume 22, Supplement December 2004, DC13–18.

Lundvall, B.-A. (1988). 'Innovation as an Interactive Process – From User-Producer Interaction to National Systems of Innovation', in G. Dosi, C. Freeman, R. Nelson, G. Silverberg, L. Soete (eds), *Technical Change and Economic Theory*, Pinter Publishers, London.

Mathews, J. A. (2000). 'Accelerated Technology Diffusion through Collaboration: The case of Taiwan's R&D Consortia', Working paper no. 106, The European Institute of Japanese Studies.

Meijer, K. F. and J. Wilting (1997). 'Trends in the Organization of Drug Research: Interfacing Industry and Universities', *Journal of Pharmaceutics and Biopharmaceutics*, Volume 43, Issue 3, June: 243–52.

Ministry of International Trade and Industry, Malaysia (2006). 'Third Industrial Master Plan 2006–2020'.

Ministry of Science, Technology and Environment Malaysia (1997). *Assessment of Biological Diversity in Malaysia*, Kuala Lumpur.

Ministry of Agriculture and Rural Development, Socialist Republic of Vietnam (2000), 'Agricultural and Rural Development 5 Year Plan (2001–2005)'.

Ministry of Agriculture and Rural Development (MARD) (2003). 'Biotechnology R&D in the sector of Agriculture', Conference Report, National Conference on Biotechnology. Hanoi, 2003.

Ministry of Education and Training (MOET). (2003). 'Training and Human Resource for Biotechnology Industry', conference report. National Conference on Biotechnology. Hanoi, 2003.

Ministry of Health. (2003). 'Application of Biotechnology in Diagnostic, Therapeutics and Pharmaceuticals', Conference Report, National Conference on Biotechnology. Hanoi, 2003.

Ministry of Science and Technology (MOST) (2003). 'Review of the Biotechnology Development in Vietnam', Conference Report, National Conference on Biotechnology. Hanoi, 2003.

Ministry of Science and Technology (MOST) (2005), 'Science and Technology Indicators', White Book. Hanoi.

MOEA (2005). 'Biotechnology Poised to be Taiwan's Next High Technology Success Story', *Biotechnica*, Biotechnology and Pharmaceuticals Industries Program Office (BPIPO), Taiwan.

Mosher, A. (1966). *Getting Agriculture Moving: Essentials for Development and Modernisation*, New York: Praeger Publications.

Motari, M., J. M. Wekunda and J. N. Kabare (2001). 'A report of the National Workshop on the Needs Assessment for the Kenyan Biotechnology and Biosafety System', 1–2 November 2001, Nairobi, Kenya.

Mugabe, J. (2002). 'Biotechnology in sub-Saharan Africa: Towards a Policy Research Agenda,' ATPS Special Paper no. 3. Nairobi, Kenya.

Mundeak, Y. (1997). 'Agricultural Production Functions: A Critical Survey', Working paper, The Centre for Agricultural Economic Research, Rehovot, Israel.

National Office of Intellectual Property (NOIP, 2005). Annual Report. Hanoi.

National Council for Science and Technology (NCST). (2005). A Guide to the Biosafety Act – 2005.

National Council for Science and Technology (NCST). (2005). National Policy on Biotechnology, 2005.

National Council for Science and Technology (NCST). (2005). The Biosafety Bill, 2005.

National Council for Science and Technology (NCST). (2004). A Manual for Inspection and Monitoring of Genetically Modified Organisms (GMOs) in Kenya.

National Council for Science and Technology (2003). 'Implementation of the National Biosafety Framework for Kenya', proceedings of the UNEP-GEF Workshop 14–18 April 2003, Nairobi, Kenya.

Nelson, R. R. and Winter, S. G. (1982). *An Evolutionary Theory of Economic Change*, Belknap Press, Cambridge, MA and London.

Nelson, R. R. and Winter, S. G. (1982b). 'The Schumpeterian Tradeoff Revisited', *American Economic Review*, Vol. 72, 114–32.

Nelson, R. (1993). *National Innovation Systems: A Comparative Analysis*. Oxford University Press. Oxford.

Nelson, R. (1996). 'What is "Commercial" and What is "Public" about Technology and What Should be?' in N. Rosenberg, R. Landau, and D. C. Mowery (eds), *Technology and the Wealth of Nations*. Stanford, CA: Stanford University Press.

Ngamau, C., B. Kanyi, J. Epila-Otara, P. Mwangingo, and S. Wakhusama (eds). (2004). *Towards Optimizing the Benefits of Clonal Forestry to Small Scale Farmers in East Africa*. ISAAA Briefs No. 33, 2004.

Ngoc Hai, N. (1998). 'Organic Agriculture in Developing Countries Needs Modern Biotechnology', *Biotechnology and Development Monitor*, No. 34, p. 24.

Ngo Luc Cuong (2004). *Assessments of the Needs in Biotechnology Applications in Least Developed Countries: Case Study in Vietnam*. UNU-IAS Working Paper No.122.

Nguyen Van Tuong (2000). 'Agricultural Biotechnology in Vietnam', paper for *Regional Conference on Agricultural Biotechnology*. Bangkok, Thailand.

Nguyen Vo Hung (2003). 'Technology Market Development', working paper. NISTPASS. Hanoi.

Nguyen Thanh Tung (2004). 'Guidance for SMEs in Technology Innovation', working paper. NISTPASS. Hanoi.

Ninth Malaysia Plan, Government Printers (2005).

NISTPASS (2003). Vietnam's Science and Technology Strategy until 2010. Hanoi.

North, D. C. (1996). 'Epilogue: Economic Performance through Time', in L. J. Alston, T. Eggertson and D. C. North (eds) *Empirical Studies in Institutional Change*, Cambridge University Press, Cambridge, pp. 342–55.

North, D. C., W. Summerhill, and B. R. Weingast (2000). 'Order, Disorder and Economic Change: Latin America vs. North America' in Bruce Bueno de Mesquita and Hilton Root (eds), *Governing for Prosperity*. Yale University Press: New Haven.

Nyameimo, D. M. (2006). 'Towards a Regional Approach to Biotechnology & Biosafety Policy in Eastern and Southern Africa: Stakeholders Analysis Report on the Implications of Modern Biotechnology on Trade and Food Security In Kenya', final study report under the RABESA initiative.

Odame, H. (2005). 'Thinking about Local Set-Ups: Making Sense of Biotechnology in Kenyan Agriculture', paper presented at the Workshop on the Globalization of Agricultural Biotechnology: Multi-disciplinary views from the South. Center for the Study of Globalization and Regionalization at the University of Warwick, 11–13 March 2005.

Odame, H., P. Kameri-Mbote and D. Wafula (2002). 'Innovation and Policy Process: case of Transgenic Sweet Potato in Kenya', *Economic and Political Weekly*, July 6, pp. 2727–77.

OECD (1998), The Emerging Digital Economy, OECD: Paris.

Okafor, N. (1994). 'Biotechnology and Sustainable Development in Sub-Saharan Africa', *World Journal of Microbiology and Biotechnology*, 10: 243–8.

OTA (1989). 'New Developments in Biotechnology: Patenting life', special report OTA-BA-370. Washington DC: US Government Printing Office, pp. 25–48.

Oyelaran-Oyeyinka, B. and Barclay, L. A. (2003) *System of Innovation and Human Capital in African Development*, UNU-INTECH Discussion Paper, Maastricht, Netherlands.

Oyelaran-Oyeyinka, B and Gehl Sampath, P, *Latecomer Development: States, Knowledge and Economic Growth*, Routledge, London, 2009.

Patel, S. (2005). 'State of Biotechnology Research and Future Trends in Kenya: GTIL Achievements and Challenges in the Field of Biotechnology', Unpublished manuscript.

Pingali, P. and T, Raney. (2005). 'From Green Revolution to the Gene Revolution: How will the Poor Fare?', ESA Working paper no. 05–09. Agriculture and Development Division. FAO.

Quan, Nguyen Manh (2004). 'Development of Biotechnology and Biotechnological Industry in Viet Nam – Emerging Issues', Interim report for NISTPASS. Hanoi.

Republic of Kenya (1926). The Crop Production and Livestock Act cap 321 Laws of Kenya. Government Printer, Nairobi.

Republic of Kenya (1965). Food, Drugs and Chemical Substances Act (Cap 254), Laws of Kenya, Government Printer, Nairobi.

Republic of Kenya (1923). Agricultural Produce (Export) Act (cap 319), Laws of Kenya, Government Printer Nairobi.

Republic of Kenya (1937). The Plant Protection Act cap 324 Laws of Kenya. Government Printer, Nairobi.

Republic of Kenya (1977). The Science and Technology Act cap 250 Laws of Kenya. Government Printer, Nairobi.

Republic of Kenya (1974). The Standards Act cap 496 Laws of Kenya. Government Printer, Nairobi.

Republic of Kenya (1990). Industrial Property (Repealed) Act Cap 509. Government Printer.

Republic of Kenya (1975). The Seeds and Plant Varieties Act cap 326 Laws of Kenya. Government Printer, Nairobi.

Republic of Kenya (1999). Environmental Management and Coordination Act No.8. Government Printer, Nairobi.

Republic of Kenya (2003). Public Health Act (cap 242), Laws of Kenya, Government Printer, Nairobi.

Republic of Kenya (1953). Veterinary Surgeons Act (cap 366), Laws of Kenya, Government Printer, Nairobi.

Republic of Kenya (1965). Animal Diseases Act (Cap 364), Laws of Kenya, Government Printer, Nairobi.

Republic of Kenya (1989). The Industrial Property Act of 1989 (cap. 509), Laws of Kenya, Government Printer, Nairobi.

Republic of Kenya (1975). Trademarks Act (Act 506), Laws of Kenya, Government Printer, Nairobi.

Republic of Kenya (2001). The Copyrights Act (cap 150), Laws of Kenya, Government Printer, Nairobi.

Rosenberg, N. (1976). *Perspectives on Technology*. Cambridge: Cambridge University Press.

Schultz, Theodore William (1964). *Transforming Traditional Agriculture*. New Haven: Yale University Press.

Spillane, C. (2002). 'Agricultural Biotechnology and Developing Countries: Proprietary Knowledge and Diffusion of Benefits', in T. Swanson (eds), *Biotechnology, Agriculture and the Developing World*. Cheltenham, UK and Northampton, MA, USA: Edward Elgar Publishing Ltd.

Tansey, G. (2002). 'Food Security, Biotechnology and Intellectual Property: Unpacking Some Issues around TRIPS', Quaker United Nations Office, Geneva, available at www.quno.org/economicissues/intellectual-property/intellectual-Links.htm.

Teubal, M. (1999). 'Towards an R & D Strategy for Israel', *The Economic Quarterly*, 46(2): 359–83.

Thailand's National Biotechnology Policy Framework 2004–2009 (2005). National Centre for Genetic Engineering & Biotechnology, National Science and Technology Development Agency.

Thompson, P. B. (1997). 'Food Biotechnology's Challenge to Cultural Integrity and Individual Consent', The Hastings Centre Report, 27(4): 34–8.

The International Institute of Tropical Agriculture (IITA) 2004 *Annual Report*.

The Raw Materials Research and Development Council (RMRDC) 2004.

The World Bank Group and IMD World Competitiveness Year Book 2004.

Tran Duy Quy (2005). 'Agricultural Biotechnology of Vietnam in Past Twenty Years of Renovation Process', *Science and Technology Journal of Agriculture and Rural Development*. July 2005.

Traynor, P. L. and H. K. Macharia (2003). 'Analysis of the Biosafety System for Biotechnology in Kenya: Application of a Conceptual Framework', ISNAR Country Report 65.

WIPO Patent Report (2007). Statistics on Worldwide Patent Activity.

Wafula, J. S. (1999) 'Agricultural Biotechnology in Kenya: Approach, Innovation and Challenges', a background paper prepared for the Regional Workshop on Biotechnology Assessment: Regimes and Experiences September 27–29, 1999, Nairobi, Kenya.

Wafula, J. and C. Falconi (1998). 'Agricultural Biotechnology Research Indicators', Kenya Discussion Paper No. 98, 9 September 1998, The Hague: ISNAR.

Wafula, J. (2000). 'Building Partnerships for Biotechnology Development and Networking in Africa', 'KARI Final Project Report', *Report submitted to USAID-MSU ABSP*, June, Nairobi: KARI.

Wagner, C. S. and M. S. Reed (2005). 'The Pillars of Progress: Metrics for Science and Technology Infrastructure', *Industrial Development Report 2005*, Background Paper Series. UNIDO.

World Development Indicators, (2007).

Websites

General Office of Statistic (GSO): www.gso.gov.vn.

http://www.isaaa.org/kc.

FAO 2005: http://faostat.fao.org.

www.biotechnology.gov.au.

NZBiotechnologyindurstry2006: http:/www.nzbio.org.nz/uploaded/NZBio_Growth_Brochure.pdf.

http://www.nzbio.org.nz/uploaded/NewZealand2005BiotechnologyArticle FinalVersion.pdf.

http://www.siliconvalley.com/mld/siliconvalley/business/industries/.

http://www.isaaa.org/kc.

http://arno.unimaas.ni/show.cgi?fid=4483.

www.grain.org/.

Index